DATE DUE

BRODART, CO.   Cat. No. 23-221-003

Innovations in Teaching
Abstract Algebra

©2002 by the Mathematical Association of America (Inc.)

ISBN: 0-88385-171-7

Library of Congress Catalog Card Number 2002101377

Printed in the United States of America

Current Printing

10 9 8 7 6 5 4 3 2 1

Innovations in Teaching Abstract Algebra

Allen C. Hibbard and Ellen J. Maycock

Editors

Published and Distributed by
The Mathematical Association of America

The MAA Notes Series, started in 1982, addresses a broad range of topics and themes of interest to all who are involved with undergraduate mathematics. The volumes in this series are readable, informative, and useful, and help the mathematical community keep up with developments of importance to mathematics.

MAA Notes

38. Models That Work: Case Studies in Effective Undergraduate Mathematics Programs, *Alan C. Tucker,* Editor.

39. Calculus: The Dynamics of Change, *CUPM Subcommittee on Calculus Reform and the First Two Years, A. Wayne Roberts,* Editor.

40. Vita Mathematica: Historical Research and Integration with Teaching, *Ronald Calinger,* Editor.

41. Geometry Turned On: Dynamic Software in Learning, Teaching, and Research, *James R. King and Doris Schattschneider,* Editors.

42. Resources for Teaching Linear Algebra, *David Carlson, Charles R. Johnson, David C. Lay, A. Duane Porter, Ann E. Watkins, William Watkins,* Editors.

43. Student Assessment in Calculus: A Report of the NSF Working Group on Assessment in Calculus, *Alan Schoenfeld,* Editor.

44. Readings in Cooperative Learning for Undergraduate Mathematics, *Ed Dubinsky, David Mathews, and Barbara E. Reynolds,* Editors.

45. Confronting the Core Curriculum: Considering Change in the Undergraduate Mathematics Major, *John A. Dossey,* Editor.

46. Women in Mathematics: Scaling the Heights, *Deborah Nolan,* Editor.

47. Exemplary Programs in Introductory College Mathematics: Innovative Programs Using Technology, *Susan Lenker,* Editor.

48. Writing in the Teaching and Learning of Mathematics, *John Meier and Thomas Rishel.*

49. Assessment Practices in Undergraduate Mathematics, *Bonnie Gold,* Editor.

50. Revolutions in Differential Equations: Exploring ODEs with Modern Technology, *Michael J. Kallaher,* Editor.

51. Using History to Teach Mathematics: An International Perspective, *Victor J. Katz,* Editor.

52. Teaching Statistics: Resources for Undergraduate Instructors, *Thomas L. Moore,* Editor.

53. Geometry at Work: Papers in Applied Geometry, *Catherine A. Gorini,* Editor.

54. Teaching First: A Guide for New Mathematicians, *Thomas W. Rishel.*

55. Cooperative Learning in Undergraduate Mathematics: Issues That Matter and Strategies That Work, *Elizabeth C. Rogers, Barbara E. Reynolds, Neil A. Davidson, and Anthony D. Thomas,* Editors.

56. Changing Calculus: A Report on Evaluation Efforts and National Impact from 1988 to 1998, *Susan L. Ganter.*

57. Learning to Teach and Teaching to Learn Mathematics: Resources for Professional Development, *Matthew Delong and Dale Winter.*

58. Fractals, Graphics, and Mathematics Education, *Benoit Mandelbrot and Michael Frame,* Editors.

59. Linear Algebra Gems: Assets for Undergraduate Mathematics, *David Carlson, Charles R. Johnson, David C. Lay, and A. Duane Porter,* Editors.

60. Innovations in Teaching Abstract Algebra, *Allen C. Hibbard and Ellen J. Maycock,* Editors.

MAA Service Center
P. O. Box 91112
Washington, DC 20090-1112
800-331-1622   fax: 301-206-9789

Preface

Over the past decade, the undergraduate abstract algebra classroom has undergone a dramatic transformation. Many faculty who were exposed to new pedagogical techniques during the calculus reform wanted to experiment with those techniques in more advanced classes. A variety of software packages were written or extended for use in abstract algebra. This collection of articles is the outgrowth of several gatherings of mathematicians who were interested in discussing the teaching and learning of abstract algebra. We, the editors of this volume, organized two contributed paper sessions entitled *Innovations in Teaching Abstract Algebra* at the 1997 and 1999 Joint Mathematics Meetings. One of the editors co-organized the NSF-UFE workshop *Exploring Undergraduate Algebra and Geometry with Technology* held on the DePauw University campus in June, 1996. We invited participants from these two contributed paper sessions and the workshop, as well as several other mathematicians, to write articles on a variety of new approaches in teaching abstract algebra.

We have chosen the articles that appear in this volume for several different purposes: to disseminate various technological innovations, to detail methods of teaching abstract algebra that engage students, and to share more general reflections on teaching an abstract algebra course. We expect that the reader of this volume will be either a faculty member who is new to the teaching of abstract algebra or a seasoned teacher of algebra who is interested in trying some new approaches. In either case, we hope that the reader will be intrigued and stimulated by these diverse expositions.

We have divided the articles into three broad, but ultimately overlapping areas: *Engaging Students in Abstract Algebra*, *Using Software to Approach Abstract Algebra*, and *Learning Algebra Through Applications and Problem Solving*. Below we give an overview of the articles in each section. Additionally, each article begins with a brief abstract. Since some of the articles rely on colored diagrams, have downloadable materials, or are best read while using some particular software, we have created a web site that accompanies this volume. It can be found at

$$\text{http://www.central.edu/MAANotes/.}$$

This site will be maintained to provide up-to-date sources for all materials, software, and web sites referenced in all of the articles. In fact, to fully appreciate some of the articles the reader may wish to visit this site for the color images that could not be captured in this black and white production.

Engaging Students in Abstract Algebra. We begin this volume with several articles that present individual overviews of course structure. While these are personal examples, we hope that readers can extract useful information for their own classrooms. After a meeting at a NExT session, Laurie Burton, sarah-marie belcastro, and Moira McDermott decided to share their experiences with each other as they each navigated teaching algebra for the first time. The reflections in their article may be particularly appropriate for other first-time algebra instructors. Steve Benson and Brad Findell provide a good discussion of some modified discovery techniques that the reader can implement. Gary Gordon uses geometry to help teach group theory, aided by several dynamic software packages. His article focuses on how the groups of symmetry can illustrate many algebraic concepts. Paul Fjelstad discusses how he has used some concrete experiences (special decks of cards, for example) to motivate students to build various abstract structures. Su Dorée shares how she used the theme of the number of solutions of $x^2 + 1 = 0$ as a student research project. She includes some guidelines to consider when incorporating such a project.

Using Software to Approach Abstract Algebra. One of the earliest to use software to teach abstract algebra was Ladnor Geissinger, who created *Exploring Small Groups* (*ESG*). This DOS-based program is the foundation for Ellen Maycock Parker's volume *Laboratory Experiences in Group Theory*. Her article illustrates how to integrate a laboratory component into an algebra classroom. Edward Keppelmann and Bayard Webb contribute an article discussing the program *Finite Group Behavior* (*FGB*) that they have created. Intended as a successor to *ESG*, *FGB* is a *Windows*-based program that is more flexible than *ESG*. The programming language *ISETL* has also made its way into algebra classes, including being the foundation for *Learning Abstract Algebra with ISETL* by Dubinsky and Leron. Ruth Berger, Karin Pringle, and Robert Smith each contributed articles, with different emphases, based on *ISETL*. These articles all provide examples of *ISETL* code and reasons for considering this language.

Although generally considered as a research tool for algebraists, *Groups, Algorithms and Programming* (*GAP*) can also be used in the classroom. Juli Rainbolt's article indicates her methods for doing this. In contrast, while some readers may not regard *MATLAB* as a natural environment for computing in abstract algebra, George Mackiw makes a case for doing so. He illustrates how matrix groups over finite fields, with computations being performed by *MATLAB*, can provide examples, problems, and opportunities for experimentation. Two other general-purpose computer algebra systems are also used for algebra. Kevin Charlwood indicates how he uses *Maple* as a vehicle for discovery in his classes. Similarly, Al Hibbard uses *Mathematica* as the programming environment to execute the *AbstractAlgebra* packages (which form the foundation for *Exploring Abstract Algebra with Mathematica*, written by Al and Ken Levasseur). With these packages, one can interactively explore most of the topics that occur in the undergraduate abstract algebra curriculum, providing a visualization of the concepts where possible. Using software to generate examples and to illustrate the abstract material has probably been the most dramatic change in the way we teach abstract algebra.

Learning Algebra Through Applications and Problem Solving. Specific problems often allow an instructor of abstract algebra to explore a concept more creatively. Included here is an article by Michael Bardzell and Kathleen Shannon describing their PascGalois project. They introduce a group-theoretic generalization of Pascal's triangle and explore some of its ramifications. In particular, the coloring provided by their accompanying software (or using the *AbstractAlgebra* packages) helps students to visualize some interesting patterns. John Wilson takes the *Lights Out* puzzle and develops a project that analyzes it from an algebraic point of view. This article also incorporates tips for those who want to include similar student projects. Similarly, John Kiltinen explores other puzzles, illustrating how various algebraic concepts (in particular, permutations, conjugates, and commutators) can be seen by studying his puzzles. In Lucy Dechéne's article, we see how group-theoretic notions even show up in change ringing (ringing bells in a prescribed fashion). She shows how British bell ringers worked with permutation groups considerably before mathematicians formalized them. An aural rather than a visual approach is yet another way to help students learn.

It has not been our intention to write the definitive volume on how to teach abstract algebra at the beginning of the twenty-first century. Indeed, students can have successful learning experiences in many different types of classrooms. Included here are only a portion of the innovations that are now being developed. We hope, however, that the ideas contained in this volume will stimulate readers to attempt some interesting experiments in their own abstract algebra classrooms. Choose a few ideas and try them out! A classroom that is active and that shows our students how creative and dynamic mathematics can be is an excellent learning environment.

Coherence in a collection of articles such as this one implies a high level of collaboration. The editors would first of all like to thank the authors of the articles for their interest in the project and their patience throughout the editing process. Special thanks go to Chris Christensen (Northern Kentucky University), Joseph Gallian (University of Minnesota Duluth), Gary Gordon (Lafayette College), Steven Hurder (University of Illinois at Chicago), Loren Larsen (St. Olaf College), and Josh Vogler (Central College student) for their careful reading and valuable suggestions. We are grateful for the insightful evaluations done by the Notes Editorial Board and especially wish to thank Sr. Barbara Reynolds for her help as we worked to prepare a final document. We appreciate the continuing support of our institutions, Central College and DePauw University. This project would not have been possible without the hard work of the folks at the Mathematical Association of America, including Don Albers, Elaine Pedreira and Beverly Ruedi. With all this support and encouragement, we have been able to create a volume that will, we think, have an important impact on the teaching of abstract algebra.

Contents

Preface vii

I. Engaging Students in Abstract Algebra

Active Learning in Abstract Algebra: An Arsenal of Techniques
Laurie Burton, sarah-marie belcastro, and Moira McDermott 3

A Modified Discovery Approach to Teaching and Learning Abstract Algebra
Steve Benson, with Brad Findell 11

On Driving Students to Abstraction
Paul Fjelstad 19

Using Geometry in Teaching Group Theory
Gary Gordon 25

An Abstract Algebra Research Project: How many solutions does $x^2 + 1 = 0$ have?
Suzanne Dorée 35

II. Using Software to Approach Abstract Algebra

Laboratory Experiences in Group Theory: A Discovery Approach
Ellen J. Maycock 41

Learning Beginning Group Theory with *Finite Group Behavior*
Edward Keppelmann, with Bayard Webb 45

Discovering Abstract Algebra with *ISETL*
Ruth I. Berger 55

Teaching Abstract Algebra with *ISETL*
Karin M. Pringle 63

Using *ISETL* and Cooperative Learning to Teach Abstract Algebra: An Instructor's View
Robert S. Smith 71

Using *GAP* in an Abstract Algebra Class
Julianne G. Rainbolt 77

Experiments with Finite Linear Groups Using *MATLAB*
George Mackiw 85

Some Uses of *Maple* in the Teaching of Modern Algebra
Kevin Charlwood 91

Using *Mathematica* to Explore Abstract Algebra
Allen C. Hibbard 97

III. Learning Algebra Through Applications and Problem Solving

The PascGalois Triangle: A Tool for Visualizing Abstract Algebra
Michael J. Bardzell and Kathleen M. Shannon 115

Developing a Student Project in Abstract Algebra: The *Lights Out* Problem
John Wilson 125

Learning Permutation Group Theory via Puzzles
John O. Kiltinen 131

Ringing the Changes: An Aural Permutation Group
Lucy Dechéne 137

Appendix

Internet Resources for this Volume 145

About the Authors 147

Index 151

Part I

Engaging Students
in
Abstract Algebra

Active Learning in Abstract Algebra: An Arsenal of Techniques

Laurie Burton, sarah-marie belcastro, and Moira McDermott

Abstract. This article describes the collaboration of three recent Ph.D.s, each teaching abstract algebra for the first time. The authors describe how they met, how they collaborated and how they addressed various pedagogical issues. In particular, they examine issues that relate to upper-division proof-writing courses. All three of the authors emphasized and encouraged active student participation in their courses. This paper provides a discussion of techniques and examples of assignments used to encourage active learning at the beginning of the course, in the classroom, outside the classroom, and through assessment. Included are notes given to students on the syllabus, some sample exploratory group activities, techniques to encourage conjecture-making and proof-building, methods for refining proof-writing, and various ways of structuring homework.

1 Introduction

Laurie, sarah-marie and Moira met as fellows of the 1997-1998 Project NExT cohort. We discovered through e-mail conversations preceding our meeting in Atlanta, Georgia (MathFest, 1997) that we were all scheduled to teach abstract algebra courses the following academic year. When we met in Atlanta we discussed various text possibilities, our plans for active presentation of course content and the structure of our courses. As a collective we decided to use Joseph Gallian's text, *Contemporary Abstract Algebra* [4] and to communicate via e-mail about our day-to-day experiences in teaching this course. We then communicated throughout the year and collaborated again, in person, at the 1998 Joint Mathematics Meetings in Baltimore and at the 1998 MathFest in Toronto. As professors in our first years of teaching, we were excited about teaching abstract algebra, and we were particularly interested in teaching our courses with an emphasis on active student participation. We shared the opinion that requiring active preparation and active involvement of ourselves would result in a greater commitment to our courses from our students. This article summarizes our discussions, collaboration and findings, explaining and giving examples of some of the teaching techniques that we found most valuable.

1.1 Our Courses and Our Students

Laurie taught two different abstract algebra courses: a summer Master's in Teaching class and a year-long, senior-level undergraduate sequence. The graduate students were highly motivated to succeed but were not uniformly strong in their undergraduate preparation. These students needed to increase their mathematical maturity and be given materials they could use in their high school classrooms. In the senior sequence, the undergraduates were still struggling with basic logic and the idea of a proof. These students needed to learn to work with advanced mathematics.

Mainly for secondary mathematics education majors, sarah-marie's course was a year-long sequence. Her students varied widely in their motivation and preparation, but most were somewhat weak. In particular, many knew nothing about the rudiments of proof, were passive learners, and expected that they would not have to spend large amounts of time on the class.

3

Moira taught a two-semester sequence aimed primarily at math majors. Her students were generally strong, although several had little or no experience writing proofs.

2 Ways To Start Students On Active Learning

2.1 Notes on Writing Proofs

As with all courses, it is important to set an enthusiastic yet hard-working tone in the classroom from the beginning of the term. We each provided our students with a syllabus that clearly outlined our expectations. To help the students realize active proof-writing would be an important focus of the course, Laurie included the following on her undergraduate course syllabus. (For similar ideas on writing pointers, see [2].)

Tests are fun, but the real focus of this course will be on successfully writing the homework. I have several books on writing mathematics and writing mathematical proofs (e.g., [7]). I will occasionally refer to them; you may borrow them to look over at any time.

Please keep in mind when you write up your homework that you should:

- Clearly (re)state the problem.
- Clearly state any assumptions you are making.
- Clearly state which proof method you are using if it is not a straightforward proof (e.g., "I will proceed by considering the contrapositive").
- Carefully show each step you are taking.
- Clearly reference any results that you are using.
- Conclude your proof.

Please also keep the following in mind when you read over your homework before turning it in:

- Are the spelling, grammar, and punctuation correct?
- Is the mathematics correct?
- Did you answer the question or prove the statement that was originally asked or given?
- Is this paper neat, organized, and pleasingly presented?

And last, but definitely not least: Read over each sentence of your proof. Try reading your proof out loud. Does it make sense? Is each sentence complete? Are any steps left out? Are there any amazing or unverified leaps? If you are using previous results, or results from the book, have you referred to them (e.g., we know this from Theorem 4.2)?

2.2 Notes on Style

Since sarah-marie was not prepared for the passivity of her students, she did not explicitly address the issue of active learning in her initial handouts. Over the course of her first term, she realized she needed to remind her students frequently that it is normal to read the book over and over. She also found it necessary to explicitly state that she expected them to spend at least two hours per week reading the book and at least eight hours per week on homework. The second term, she added the following to her first-day handout:

You can think of this as a seminar course if you would like a pigeonhole to put it in. There will be at most one third of each class spent on lecturing, and in reality, the average amount of lecture for the first half of the course will be about five minutes per class. Most of the class will be spent answering your questions and working on problems. I view classroom time not as a time when you do most of your learning, but as a focus time where you get unstuck, or a boost, or a burst of insight. Most of your learning will take place outside the classroom, on your own time.

The students who continued from her previous class helped reinforce this paradigm for the new students.

2.3   Activity for the First Day

In order to emphasize the point that class would not be just lectures, Moira had her students work on a project the first day of class.

Students were given cardboard triangles and squares with the corners numbered. They were first asked to determine all of the symmetries of the figures. They were given instructions on how to do this based on the information about symmetries in Chapter 1 of [4]. They then came up with labels for the symmetries, for example, "rotation by 90 degrees." Moira drew a grid on the board and had each student fill in one column of the multiplication table. The students were then asked to come up with interesting properties of the table. This led to a discussion that introduced the ideas of closure, commutativity, identities, and inverses.

This activity served two purposes. First, it forced the students to do something other than sitting passively on the first day. Second, it previewed basic ideas that came up in the beginning of the course and gave students a concrete example to which they could refer as the course progressed.

3   Encouraging Participation: Active Students Inside the Classroom

Of course just suggesting to students that they need to actively participate in the course and write coherent proofs does not create a dynamic classroom and skilled proof writers. We each had our students regularly do group work in class. Group activities varied from working on selected text problems, exploratory exercises, and peer analysis of proof to completing instructor-designed worksheets. By facilitating group activities, we encouraged our students to explore the current material and helped them increase their proof-writing skills. It was clear the intensive interaction with and between our students in the classroom added to the overall successes of our courses. We found many students who were not willing to speak up in a large classroom were willing to speak up in smaller groups. We also noted that the comfortable environment of group work helped some students feel less intimidated about seeking help during office hours. In fact, throughout our courses, we all saw students who worked together in class get to know each other. This helped our students to form valuable study groups outside of the classroom.

The mechanics of managing collaborative group activities in an abstract algebra class are almost identical to those for a calculus class. We recommend [1], [6], and [3] to the interested reader.

Here are some examples of techniques we used to promote participation in the classroom.

3.1   Avoiding Lecturing

In order to help the students successfully learn to construct proofs, sarah-marie forced her students to read the text carefully. She often began class by asking students whether they had gotten stuck in the reading, and if so, where. Upon receiving the expected silence, she would ask the students to break into groups (average size of 4) and work on two or three very simple problems. Invariably, the students were quickly stuck. She would move from group to group, asking for definitions or statements of related theorems. As the course progressed, this situation changed from sarah-marie repeatedly reminding the students to look up definitions, to the students automatically analyzing the details of definitions in order to begin short proofs. After (and *only* after) students had produced and discussed relevant definitions would sarah-marie provide assistance in the stickier points of the problems.

Examples of good simple exercises to ask the students to do in groups are

- Prove that S_4 is not isomorphic to D_{12} (from the chapter on isomorphisms).

- If A and B are ideals of a ring, show that the *sum* of A and B, $A + B = \{a + b \mid a \in A, b \in B\}$, is an ideal (from the "Ideals and Factor Rings" chapter).

These problems force the students to analyze the details of a definition or several definitions; here they must use the conditions given in the definitions of isomorphism and ideal in order to complete the proof. The students must also review the structural facts they have learned about groups such as S_4 and D_{12}, which then forces them to recall concepts such as being abelian.

This technique was very successful in impressing on the students that proofs do not materialize out of thin air. They learned that they needed to remind themselves of details constantly, and that it can take a very long time to produce a proof of even a simple exercise.

3.2   Exploratory Exercises

When Moira lectured she wasn't confident her students would always ask questions about the concepts they didn't understand. Further, she worried that they might not have the skills or be confident enough to make conjectures. To address this, Moira wrote exploratory exercises for in-class group work. For example, she had her students construct groups of small order. Students were led through a series of steps that allowed them to construct the group of order three and conclude that there is only one, up to isomorphism. They also constructed two groups of order four and concluded that those were the only two. Another exercise involved determining when $U(n)$, the group of units of $\mathbb{Z}_n$, is cyclic. Students determined whether $U(n)$ was cyclic for small values of n. They were then given the information for larger values of n and asked to form a conjecture about the prime power decomposition of integers n for which $U(n)$ is cyclic. The students were allowed to ask Moira for further data and, in fact, they were able to prove their conjectures at a later point in the course. The design of this exploratory exercise was motivated by one of the Programming Exercises in the text [4, Chapter 4, #1]. In general the programming exercises in [4] are a good source of ideas for developing one's own exploratory exercises. One way for students to get data without actually doing the programming is to use the online applets found at the author's web page at the University of Minnesota, Duluth [5]. Students liked the exploratory exercises; for many of them these exercises provided their first exposure to making conjectures.

3.3   In-Class Analysis of Proofs

Laurie and Moira regularly had their undergraduates write proofs on the board. The students found discussing the material with their peers to be helpful and enjoyable. We found when students had to explain to their peers precisely what they were trying to communicate, they were better able to focus on the logical and grammatical flaws in their proofs. Furthermore, in the long run, this activity greatly helped the students become engaged in learning the material in the course and helped build the students' proof-writing skills and style.

We also found that occasionally allowing the students to write their proofs on the board for peer analysis (before the homework was due) positively affected not only student attitude and performance but also their involvement in the course.

Another technique that Moira tried was to lead her students through an analysis and synthesis of a longer proof from the book. Moira did this by:

- Photocopying a proof from the text.

- Cutting the proof into pieces.

- Taping the pieces of the proof on a piece of paper, leaving space between the pieces.

- Giving each student two copies of the cut-and-paste page.

The students were then requested to fill in the details and missing steps on one copy (analysis) and to fill in the big picture and outline the structure of the proof on the second copy (synthesis). The students were then required to do a similar analysis and synthesis exercise with another proof from the book, this time creating their own photocopied cut-and-pasted proofs.

4   Encouraging Participation: Active Students Outside The Classroom

Although we all would have liked to assume that the dynamic state of work and discussion in our classroom naturally carried into the homes and study groups of our students, we felt it would be beneficial to provide motivation for working outside of the classroom. The following outlines some of our techniques to inspire our students to be active learners at all times.

4.1   Collaborative Homework

Moira required her students to work in groups on the homework. Students were divided into groups of three or four by the instructor and these groups were changed several times during the semester. Students worked together on the problems and then divided up the problems and shared the writing of solutions. For each homework assignment, each group turned in one final version of their solutions. Additionally, each problem set contained one or two starred problems that each student had to write up independently. Moira found that the students actually did share the work. She also found the students felt less pressure than if they had been required to do all of the problems by themselves and that they thought the writing burden was less onerous when assigned in this style. It is worth noting that Moira did think that some of her students would have benefited from more individual work.

4.2   Student Papers

Each term sarah-marie required a paper, with some criteria in common. The students were allowed to suggest topics.

In the first term, she required a 2–3 page word-processed expository paper on a student-selected subject from the *Special Topics* section of [4]. The audience of the paper was specified to be the students' class peers, and the topic had to be specified a month before the paper was due. Rough drafts were requested, but not required.

As it turned out, the students were quite inexperienced at writing mathematically and most submitted several rough drafts for comments. Some of them made corrections simply because they were requested, rather than because they understood that this would improve the paper. Even so, students who continued to the next term turned in more pleasingly-written homework and exams. They had improved their proof-writing skills (some dramatically) and general mathematical communication skills through this paper.

In the second term, sarah-marie modified the instructions. A rough draft was required, significantly before the final draft was due, to prevent students from doing shoddy work at the last minute. Furthermore, each student had to give a 15–20 minute class presentation on the 5–7 page paper. Additionally, they were instructed to create a handout to be given to the class at least one period prior to the presentation. Giving the presentations served three purposes: (1) to encourage the students to revise the mathematics in their papers meaningfully rather than mechanically (2) to give the students practice presenting mathematics, and (3) to give the students practice in distilling complex ideas for a general audience. Finally, the students had to write sample problems and solutions for the final exam based on their respective topics.

The second-term papers turned out better on the whole, with fewer rough drafts needed. At this point, the students were concentrating on learning the mathematics and presenting it well.

Moira also had her students write papers during the second semester. Her paper assignment differed from sarah-marie's in that each student selected a different topic, chose reading material for the class from recommended library books, and eventually presented material on the topic to the class for two to three days. The students were generally creative in their class presentations. In addition to interactive lecturing, they created worksheets, group activities, and computer exercises for the class. The papers were an outgrowth of the presentations and they were uniformly of good quality, in part because the students had already spent a good deal of time learning the material and preparing their class presentations. Most students submitted drafts of their papers. Their ability to write about mathematics was noticeably better than it had been at the beginning of the semester.

4.3   Term Projects

Laurie wanted her graduate students to connect the abstractions of modern algebra to activities and ideas they could bring into their high school classrooms. The graduate students were instructed to imagine they were creating a special topics module (with unlimited time) for a group of promising high school students.

The term project assignment was:

Write a series of lesson plans, lecture notes, and worksheets:

- To guide your students in using manipulatives (triangle, square, and so on) to discover the properties of a group.
- To teach your students the definition of a group.
- To teach your students the very basics of modular arithmetic.
- To use the properties of a clock to model the behavior of $\mathbb{Z}_{12}$ (in the form of a worksheet).
- To teach your students enough additional information to support the following project.

Project for the (high school) students: Gather at least six items with UPC barcodes on them, and then, using the information we have learned in class, verify that the check digit is correct.

The graduate students turned in exceptional term projects. They fully embraced the ideas behind the assignment and utilized their skills as mathematics teachers to synthesize some of the ideas of their high-level course into clear concepts that they could share with their students.

5   Active Assessment

We all felt homework should be both designed and graded to foster quality proof-writing skills while also providing general encouragement to our students. We used widely varying techniques to achieve these goals.

5.1   Copious Homework Comments

In contrast to proof-writing on the board and in-class peer review, sarah-marie's method for getting her students to write good proofs consisted entirely of working with their homework. She wrote numerous comments on their wording and grammar, and on whether statements were missing or unnecessary. This resulted, in some cases, in returned homework containing more of sarah-marie's writing than the student's. She regularly provided her students with homework solutions so they would have examples to work from and improve upon. When stopping by the Math Lab, sarah-marie often found her students discussing which points were important to make in their proofs and how to phrase them so that they would be acceptable.

Rewrites on the homework were not allowed by sarah-marie; she did this in order to encourage the students to try quite hard to turn in quality work. She found that one of the results of her strict grading policies was that many of the students asked her to look over their proofs before they handed them in to be graded.

5.2   Rewriting Homework

Laurie is a firm believer in the value of students rewriting selected homework problems until they are correct. She utilized this technique in both her graduate and undergraduate courses. Her grading procedure was simple; she would include brief but helpful comments by the homework problems (mostly on the proofs) that were poorly or illogically presented. The students were requested to rewrite the problem(s) and resubmit the corrected paper (for a reduced score on the resubmitted problems). Most of the students were able to clearly understand, with a few comments, what their errors were, and they were enthusiastic for the opportunity to rewrite the homework problems that they had originally missed. Laurie found the rewriting requirement to be a valuable aid in the instruction of her students; it helped them to improve their proof-writing skills and it helped their retention of core material for the exams.

5.3   Take-home Exams

Because writing proofs was an important part of all three of our courses, we all thought that is was important that our exams have some take-home component. Moira had originally planned to give in-class exams as well as take-home exams. In the end, we all gave only open-book, take-home exams. Moira wanted to ask her students to prove things on the exams and didn't want time to be an issue. Laurie's exams were mostly book problems without hints. Her intent was that the questions would be fairly easy and accessible if the students had been paying attention and doing the homework. She found the Supplementary Exercises in [4] to be a good source of exam questions. In contrast, sarah-marie gave more difficult exams, picking the hardest problems from the book. Her students spent forty to fifty hours on each exam, of which about ten were spent in sarah-marie's office. It was an enormous time commitment, but both sarah-marie and her students thought they learned a tremendous amount from the exams.

6   Our Collaboration

From the beginning we found our collaboration to be supportive and inspirational. At the 1997 MathFest in Atlanta we were able to choose [4] as our text and actually confer with the author regarding our ideas for our courses. This helped us to feel like we were part of a strong collaborative effort instead of only feeling like less experienced individuals. Our discussions helped us to refine our course designs and gain useful feedback regarding the practicality and usefulness of our specific ideas. Throughout the year we asked each other a lot of questions, both on content details and on how we planned to present different concepts. These discussions were valuable when one member of our team had strength in a particular topic and generated useful ideas for all of us.

7 Active Professors

We each endeavored to organize our courses to elicit maximum participation, to emphasize active learning, and to create an environment in which our students worked hard but found the work rewarding and enjoyable. In order to accomplish these goals, we found it necessary to be active professors, to exemplify for our students an active commitment to our abstract algebra course. As is shown by our various in-class and out-of-class techniques, we all provided a great deal of feedback both in our classrooms and in response to student work. Furthermore, we tried to do this with an enthusiastic and positive attitude. As a result, students spent a lot of time in our offices analyzing the course material with us. Overall, we found the extra encouragement and time we provided for our students produced effective courses focused on student success.

References

[1] T. Angelo, and P. Cross, *Classroom Assessment Techniques*, 2nd ed., Jossey-Bass Publishers, San Francisco, 1993.

[2] A. Crannell, `http://www.fandm.edu/Departments/Mathematics/writing_in_math/writing_index.html`.

[3] E. Dubinsky, D. Mathews, and B. Reynolds (Editors), *Readings in Cooperative Learning for Undergraduate Mathematics*, MAA Notes 44, Mathematical Association of America, Washington, DC, 1997.

[4] J. Gallian, *Contemporary Abstract Algebra*, 4th ed., Houghton Mifflin, Boston, 1998.

[5] J. Gallian, web site for applets, `http://www.d.umn.edu/~jgallian/msproject/project_head.html`.

[6] N. Hagelgans, B. Reynolds, K. Schwingendorf, D. Vidakovic, E. Dubinsky, M. Shahin, and J. Wimbish, *A Practical Guide to Cooperative Learning in Collegiate Mathematics*, MAA Notes 37, Mathematical Association of America, Washington DC, 1995.

[7] D. Solow, *How to Read and Do Proofs*, John Wiley and Sons, Inc., New York, 1990.

Laurie Burton, Mathematics Department, Western Oregon University, Monmouth, Oregon 97361; `burtonl@wou.edu`; `http://www.wou.edu/las/natsci_math/math/burton/burton.html`.

sarah-marie belcastro, Department of Mathematics, University of Northern Iowa, Cedar Falls, IA 50614-0506; `smbelcas@math.uni.edu`; `http://www.math.uni.edu/~smbelcas/`.

Moira A. McDermott, Department of Mathematics and Computer Science, Gustavus Adolphus College, St. Peter, MN 56082; `mmcdermo@gac.edu`; `http://www.gac.edu/~mmcdermo`.

A Modified Discovery Approach to Teaching and Learning Abstract Algebra

Steve Benson, with Brad Findell

Abstract. In this article, we describe an abstract algebra class taught at the University of New Hampshire. Rather than the traditional lecture/discussion model, the course was taught using a modified discovery approach in which students worked in cooperative groups. Activities were designed to encourage students to think through, debate, and come to a consensus about the essential concepts and methods of group theory. With this approach, the instructor was able to observe students working in class, ascertain their understanding, and plan subsequent lessons based on these observations. In addition to a general overview of the course, specific significant events are described and some suggestions for future implementations are provided.

1   Introduction

> *Mathematics is like a video game;*
> *If you just sit and watch,*
> *You're wasting your quarter (and semester).*

For years I have used the above saying to explain to my students why I assign and grade homework in my classes. As the years went by, I began to ask myself whether I really believed this. If I did, then why did I spend so much time in front of the class *showing* my students how I do mathematics? I tried to encourage my students to actively participate in class discussions, to make suggestions as we solved problems, and to ask questions. But I realized that I was too willing to "take over" discussions and too quick to do so. I had worried about these issues before but each time went back to the teaching style with which I was most comfortable and successful.

Something else bothered me as well. If a student could not solve a problem or prove a theorem on a test, did that mean the student did not understand important concepts or did it mean the student had trouble with that particular problem at that particular time? Alternatively, if a student did well on a test, did that mean the student had a good understanding of the material or did it merely mean the student was able to solve the particular problems on the test? I started to assign take-home exams and quizzes. Although I was not thrilled with everything I was doing, I could not think of any better ways to assess their understanding.

One day, at the end of a class for pre-service secondary teachers, a student looked up from her desk and said, "Wow, you sure know a lot of stuff!" I thought I had chosen activities that would help my students see that *they* could solve problems that were new to them. This young woman thought I just remembered how to solve the problems—even though the problems were new to me too. It finally occurred to me that if one learns mathematics by doing mathematics (as I claimed), then perhaps the only person learning when I am up at the board is me. What should I do?

While I was assigned to teach an introductory abstract algebra course at the University of New Hampshire, Brad Findell (a graduate student in the Department of Mathematics Education at the time) was studying how students learn concepts of abstract algebra. Brad and I found that we shared a constructivist viewpoint of learning. Here is a brief introduction to the philosophy of constructivism [4]:

> Constructivism takes the position that knowledge is not passively received, but is actively built up by the learner. ... Individuals actively try to make sense of experiences within their existing ways of thinking.

We noted that there was much discussion about improving the learning of abstract algebra, but scant research. We found that "the learning in abstract algebra courses is less than satisfactory" and much of what had been written was "not strictly research, but suggestions for teaching or curriculum" [4]. Other than anecdotal data provided by instructors, there was not much information on how student learning and understanding was affected by teaching innovations. We hoped that by combining forces we could add to the research literature and provide curricular recommendations. Brad's dissertation [5] adds to the research discussion; we hope that this article will provide pedagogical insights.

2   Description of the Course

The course was in some ways similar to those I had taught in the past. I required a standard introductory text ([7]), assigned and graded regular homework, gave exams and quizzes, and followed the usual topics. What was different about this course was that I occasionally used Ladnor Geissinger's *Exploring Small Groups* software [8]. More importantly, this time my *students* covered the material as they worked through problem sets that Brad and I designed.

The course was planned and taught with one basic philosophy: By observing the students as they worked on problems, Brad and I could better judge which concepts they understood (and could apply in other contexts) and which concepts needed more work. This was a major departure from my previous classes. In the past, my understanding of what my students knew was often an assumption based solely on what I had done in class rather than what they actively displayed. If I covered material in class, so had they. I could move on to the next topic. Other than nods in class, which I took to represent understanding, I did not know whether they "got it." While it was my hope that I could transfer my understanding of a topic to my students through a carefully planned lecture or class discussion, it is not that simple. It does not matter what I know; it is what the students know that is important.

The class was designed to give us more accurate and immediate feedback from the students. This enabled us to alter the course in reaction to students' (apparent) understanding. We first had to free ourselves from the bonds of a course calendar that mapped out every class period. Before the first class, Brad and I met to discuss what we hoped the students would learn and what skills we wanted them to develop. As the semester progressed, we continued to meet frequently to compare notes on the progress of the students and to plan the next activities. As I thought about class activities, I had to force myself to think in a new way. I had to get over the thought that "teaching is telling." By spending class time watching students work on problems and talk about mathematics, I could more accurately gauge their understanding. My observations allowed me to plan subsequent class activities and to change directions in response to the needs of the students during a class period.

2.1   The First Day

On the first day of class, we distributed an activity sheet that introduced modular arithmetic and set notation. The students worked on the problems while Brad and I observed and listened to the group discussions. We answered most of their questions with "What do you think?" or "What have you tried?" They came to realize that we really did expect them to work on problems even if we had not done an example in class.

2.2   A Defining Moment for the Class

For several class periods, the class worked on problems intended to inch them closer to the concept of a group. For example, we asked them to solve problems such as those in Table 1. These problems were designed to encourage students to reflect on their assumptions about solving equations and to help them realize the importance of the existence of additive and multiplicative inverses. Then something happened. We noticed that different conceptions

$$2x = 4 \qquad\qquad 3 + x = 4 \qquad\qquad 2x \equiv 4 \pmod 5$$
$$2 + x \equiv 4 \pmod 5 \qquad 3x \equiv 4 \pmod 5 \qquad 3x \equiv 6 \pmod 8$$
$$3x \equiv 5 \pmod 8 \qquad 3x \equiv 4 \pmod 8 \qquad 2x \equiv 4 \pmod 8$$

Table 1: Problems to Solve.

(and misconceptions) were developing. For example, some students insisted that $a \equiv b \pmod{n}$ implied a was less than b, while others claimed b must be less than a. Some said b had to be the remainder when a was divided by n, while others insisted a was the remainder when b was divided by n. Brad and I met to discuss their interpretations, since these misunderstandings had also occurred in previous classes. We did not believe that we should just decree the interpretation. We wanted them to see that they were part of a mathematical community in which notation and terminology were determined by the community; therefore, the class needed to come to some agreement if they were going to be able to discuss the concept of congruence with some modulus.

First, we asked each group to spend about 10 minutes debating among themselves until they came to a consensus. Specifically, they were asked to write down every correct statement they could think of that was a consequence of $a \equiv b \pmod{n}$. Then, they were asked to share their conclusions with another group, who in turn were to ask clarifying questions and to decide whether the two groups were in consensus. There tended to be general agreement within a group, because by then they had been working together for about a week. When groups began sharing their findings with other groups, the discussions became quite heated and there was not much common ground. At this point, Brad and I decided to have a whole-class discussion. We asked for consequences of $a \equiv b \pmod{n}$ and wrote them on the board, whether they were correct or not. We thought this airing of differences in interpretation was exactly what we needed and were happy with how things were going, but slowly the atmosphere of the room began to move from a debate over interpretation to anger over the fact that Brad and I would not settle the issue. We did not want to slide back into the expert mode, since we wanted them to see the value of the discovery approach and to become less dependent on us as arbiters of mathematical truth. After much soul searching, I responded (more or less) as follows:

> There's a lot of really good discussion going on here, but I can sense some frustration. In fact, I've heard several people ask "Why don't they just *tell* us?" Now, I'm not trying to be a smart aleck, so don't take this the wrong way. The truth is, we *did* tell you ... on the first day of class! The definition of congruence (mod n) was given on the worksheet distributed the first day of class. This is one of the reasons we're teaching the class in this manner—it doesn't necessarily matter what Brad or I *tell* you. What matters is what you understand and how you interpret notation and terminology. Of course, it would be nice if we all agreed, but we don't all think alike. By talking to each other and discussing our work, we all come to a better understanding of the course material.

Fortunately, the students saw both the humor and the relevance of this situation and for several weeks, there were no complaints about the approach we were taking. However, we found it important to address periodically any frustrations caused by the practice of letting them work it out for themselves while we only facilitated. Therefore, we had occasional "summarizing days," during which we came together to discuss the big picture and review what we knew and were able to do.

It is important to mention here that the frustration level described did not occur very often. We will go into more detail later, but most of the students eventually acknowledged that the approach did help them learn.

2.3 Use of the Text

For a typical class, we constructed worksheets consisting of problems (typically propositions to be proven or calculations to be made) along with relevant definitions and a discussion of various concepts. The majority of the students' time was spent working on the problems in self-selected groups containing 3 or 4 students. Homework was assigned and collected regularly. Even though the students had worked on many of the problems in class, we still wanted them to write up their proofs and solutions carefully; learning to write mathematics was one of the major goals we had for the course. Although we occasionally asked the students to read specific sections of the text, the main reason we required a book was for the comfort of the student. It is fascinating to note, however, that with this class, we were asked significantly more questions concerning statements or problems in the textbook than we had ever received from students in any previous class. Perhaps in a typical lecture class students assume they know the material because the teacher tells them everything they need to know, so reading the book seems redundant.

2.4 The Classroom as a Mathematical Community

During the semester, when anyone had a conjecture, Brad or I made note of it on the board and alerted the class. These conjectures helped determine the direction of the next several class periods, because we incorporated them into

$+_4$	0	1	2	3
0	0	1	2	3
1	1	2	3	0
2	2	3	0	1
3	3	0	1	2

$*$	e	a	b	c
e	e	a	b	c
a	a	b	c	e
b	b	c	e	a
c	c	e	a	b

Table 2: Two Cayley Tables.

subsequent worksheets. In particular, some early conjectures concerned the division algorithm. One student even recognized that the set of linear combinations of two positive integers seemed to be the multiples of their greatest common divisor. Although the majority of class time was spent with students working on problems in groups, we occasionally had summarizing days, as described in section 2.2. We found it useful to bring the class together when an individual or group had an interesting observation or question. Such sessions were useful to encourage the class to come to a consensus concerning terminology or notation. For example, the class determined the "direction" of the permutation operation after discussing the perceived advantages and disadvantages of both choices. It was ultimately decided that the permutation $\alpha\beta$ should represent the function composition $\alpha \circ \beta$ (read from right to left), in large part because of students' comfort with functional notation. We did, however, insist that each student use the same convention so we could all talk to each other about our work and observations without having to translate.

A benefit of this approach was the fact that the students began to see the classroom as a mathematical community. Norms for notation, vocabulary, and terminology were determined by the community. Sometimes the choices made were different from those of the professional mathematical community. On these occasions, we would accept the students' choices but also informed them of the standard choice(s). For instance, a young woman in the class suggested that the groups whose Cayley tables are in Table 2 "look the same, but they're not the same ..., it's like they're *congruent*." The class quickly adopted her terminology. Of course, we told them that the standard term was *isomorphic* but encouraged them to continue to use *congruent* during the course. Thinking of isomorphic groups as congruent seemed to benefit the subsequent discussion of isomorphism. When we discussed congruent polygons, we first described the correspondence between vertices, which led to a natural reason for identifying corresponding group elements. Thus, the class created their own natural bridge between the concept of two groups being congruent (by virtue of the fact that their Cayley tables were identical) to the existence of a bijective, Cayley table-preserving function from one group to the other.

2.5   Discovering the Group Laws

Brad and I struggled with how to develop the concept of a group in a way that did not simply involve us listing the properties of a group. We began by asking the students to solve a number of congruences. They quickly recognized that some of these congruences had a unique solution, while others had either several solutions or no solutions at all. They also noticed that difficulties occurred when they attempted to solve congruences such as $2x \equiv 4 \pmod 6$ and $2x \equiv 5 \pmod 6$, where the modulus and the coefficient of x were not relatively prime. This led to a discussion of why we can divide both sides of an equation by 2 when solving equations such as $2x = 4$ (and preserve the equality) but dividing $2x \equiv 4 \pmod 6$ by 2 gives us the solution $x \equiv 2 \pmod 6$ but not the other solution $x \equiv 5 \pmod 6$. Another student recalled the Cancellation Law of the integers. This led to a discussion of when cancellation can be applied in solving congruences. They found that it worked when solving $2x \equiv 4 \pmod 5$, but not when solving $2x \equiv 4 \pmod 6$.

While constructing a variety of multiplication tables for $\mathbb{Z}_n$, the students began to make connections between invertibility and those numbers that were not zero divisors. Some students suggested reducing the multiplication tables by including only those elements that had a 1 in both its row and column. This led to a discussion of the connection between these elements and elements of the set $\{a \in \mathbb{Z}_n \mid \exists\, x \in \mathbb{Z}_n, ax = 1\}$. They had recognized the importance of an identity element and the existence of inverses in solving equations of type $ax = b$. Concerning associativity, I worried that I would have to just tell them about the property, since I did not think it would be obvious to them. However, thanks to *Exploring Small Groups* (ESG), I never had to do that.

To discover associativity, I had the students consider the *commutative loop*, one of the operation tables stored in *ESG*, whose table is given in Table 3. I asked the class to use the table to solve the equation $3x = 5$. As I had hoped, some students noticed that $3 * 6 = 5$ while others used the fact that 3 is its own inverse to solve as follows:

*	1	2	3	4	5	6
1	1	2	3	4	5	6
2	2	3	4	5	6	1
3	3	4	1	6	2	5
4	4	5	6	1	3	2
5	5	6	2	3	1	4
6	6	1	5	2	4	3

Table 3: Commutative Loop.

$3x = 5 \Rightarrow 3*3*x = 3*5 \Rightarrow x = 2$. A discussion ensued about which solution was correct and someone eventually noticed that, in fact, $x = 2$ is *not* a solution, since $3*2 = 4 \neq 5$. I asked what, "What went wrong?" and expected to get a lot of blank stares. I was saved when a student in the back row said, "It's not associative." Following up on this statement, the rest of the class period was spent parsing out the second "solution" and determining that it is valid only if $*$ is an associative operation. Problem solved! Associativity of the operation had been deemed necessary. We added it to the identity, inverse, and closure properties that we had already identified. We were now ready for the definition of a group.

While it may seem an inefficient use of time to allow students the time to discover concepts, I found that many topics took less time than when I used a more traditional lecture/discussion approach. A case in point was the section on factor groups.

2.6 Factor groups

The concept of factor groups had always been a trouble spot in previous classes and I had spent a lot of time thinking about it. (For a more thorough description of my approach to factor groups, see [1].) In a nutshell, if H is a subgroup of a group G, I think it is important for students to view the multiplication of cosets as inheriting the operation of G. That is, if a and b are elements of G, then

$$aH * bH = \{xy \mid x \in aH, y \in bH\} = \{ah_1bh_2 \mid h_1, h_2 \in H\}.$$

Early on in the semester, we gave the students a number of problems that required them to think about sets of the form $n\mathbb{Z}$ and $n + \mathbb{Z}$. It was therefore an easy transition when we considered cosets to ask them to compute products of sets like $2 + 4\mathbb{Z}$ and $3 + 4\mathbb{Z}$.

3 Student Reaction

It would be incorrect to say all my students bought into this approach from the beginning. Early in the term, a few complained that they should not be expected to work on problems that they had not already been told how to do. Another common comment was "What's the deal with working in groups? We're doing it in all of my classes!" That I was not alone in implementing cooperative learning methods played an important role in student acceptance of my approach. Most of the students eventually appreciated the approach. Here are some student comments from teaching evaluations administered at the end of the semester.

> *Working with someone else forces me to make sure of what I'm saying and makes me defend what I'm saying.*

> *...working through problems is much more revealing than watching a professor work, but sometimes a lecture would be helpful.*

> *The activities were definitely useful and helped quite a bit. The group discussions and activities allowed us to explore and practice ideas. I definitely think that I learned a lot more by working in groups and the overall structure of the class. The time spent in groups allowed us to be a part of the learning process.*

> *It makes it much easier to remember the material when you come up with the conjectures yourself.*

Of course, not all was rosy, but I am pleased to say that the vast majority of negative comments were not caused by philosophical differences.

4   Steve's Reflections

Overall, I was very pleased with the way things went. One student commented that I should try to spend an equal amount of time with every group. While this is not always practical (different groups will struggle or excel with different material), it is important to visit each group regularly. I often found it difficult to spend quality time with every group, since there were days when I spent much of the class period with one or two groups of students who were struggling with the material. In the future, I will try to at least check on everyone, both to see if there are any problems and to show everyone that I am interested in how they are doing.

Watching students doing mathematics can be an eye-opening experience. In particular, I was shocked by some of the things I heard my students say. Their misconceptions and errors can be disconcerting and it was difficult to keep myself from jumping in and correcting mistakes. I found that by being patient and allowing a classmate to correct or question an error (or for a student to discover one's own error), my students became more comfortable with their groups and more likely to speak up and contribute to group discussions.

Some people might argue that the misconceptions and errors made by students when working cooperatively is evidence that this approach leads to mistakes, but I believe just the opposite is true. Misconceptions and errors are also made in lecture/discussion classes; instructors, however, do not have the immediate feedback to quickly remedy the situation.

I also found that students in this course showed better facility with the language of algebra. I think some of this is because they were required to communicate mathematically during class time. When listening to a lecture, students are certainly exposed to ideas, but they are often not practicing communicating these ideas. In addition, I found less evidence of mindless symbol manipulation in proofs. In previous courses, students would often write proofs on homework or tests that looked like proofs I had done in class, even when the proof I had done had little or nothing to do with the proposition the student was trying to prove. I believe this happens when a mathematical proof is viewed as a game of symbol manipulation rather than as a method of justification. I conjecture that since much of the purpose for proofs in this class was for self-verification, students were less likely to present a proof they did not understand.

One final observation seems a bit paradoxical. During this course, my students were able to learn more course content than in more traditional courses I had taught. I believe this was because observing my students enabled me to more accurately assess their understanding and adjust the amount of time spent on topics.

I would be remiss if I failed to acknowledge that teaching the course in this manner took a lot of time to plan and create the worksheets. While the worksheets can be adapted to subsequent courses, it is important to realize that if the worksheets are truly designed to help students get from where they are to where one wants them to be, they cannot be created too far in advance. (For sample problem sets, quizzes, and exams, refer to this article at the web site for this volume.)

5   Suggestions

In this final section, we provide some recommendations for those who are interested in teaching in this style.

- **Require attendance.** Because a significant portion of student learning is done in the classroom, reading the text or a classmate's notes does not provide the same experience; attendance and active participation are essential. In fact, most of the students said that absences were a hardship.

- **Schedule at least one class discussion per week.** More days for summarization, especially toward the end of the semester, would be helpful.

- **Try to ensure an even pace throughout the semester.** In particular, do not try to cram in a lot at the end just to say it was covered.

- **Do more assessments.** Although assessments take a lot of time to prepare and grade, we think more assessments are important, whether using portfolios, presentations, or oral examinations. It is important to encourage student feedback early and often.

- **Use more technology.** We think we could make better use of *Exploring Small Groups*, now that Ellen Maycock Parker's laboratory manual [9] has been published. We also highly recommend considering some of the suggestions found elsewhere in this volume for integrating the computer into an algebra course.

While this approach required a lot of planning and adaptation throughout the term, I was very happy with the results. I have used this approach in a course for high school teachers and was pleased with the results. I have tried to incorporate this style into non-algebra courses. I highly recommend this style, but I close this collection of suggestions with a warning: Do not try to do too much the first time out. This was not the first time I have tried alternative teaching styles. I had to be careful not to inundate myself, since if I am overwhelmed, my students will be too. Pick some key changes to make and then stick to the decision long enough to make a realistic assessment of their effectiveness. One can always do more next time. For more suggestions on cooperative learning, see [2, 3, 6].

As a final summary of the philosophy of teaching and learning that drove us during this course, we give the last words to an anonymous Native American author:

> *Tell me and I'll forget;*
> *Show me and I may not remember;*
> *Involve me and I'll understand.*

References

[1] S. Benson and M. Richey, *Much ado about coset multiplication*, PRIMUS, **4** (1994), 9–14.

[2] N. Davidson, (Editor), *Cooperative learning in mathematics*, New York, Addison-Wesley, 1990.

[3] E. Dubinsky, D. Mathews, and B. Reynolds, (Editors), *Readings in Cooperative Learning in Mathematics, MAA Notes no. 44*, Washington, DC, Mathematical Association of America, 1997.

[4] B. Findell, *Learning in abstract algebra*, Doctoral dissertation proposal, University of New Hampshire, 1996.

[5] B. Findell, *Learning and understanding in abstract algebra*, Unpublished doctoral dissertation, University of New Hampshire, 2001.

[6] D. Finkel, and G. Monk, *Teachers and learning groups: The dissolution of the Atlas complex*, in C. Bouton and R. Garth, (Editors), Learning in Groups: New Directions in Teaching and Learning, n. 14, San Francisco, Jossey Bass., 1983

[7] J. Gallian, *Contemporary Abstract Algebra*, 4th ed., Houghton Mifflin, Boston, 1998.

[8] L. Geissinger, *Exploring Small Groups: A Tool for Learning Abstract Algebra*, now only available bundled with [9].

[9] E. Parker, *Laboratory Experiences in Group Theory: A Manual to be used with Exploring Small Groups*, Mathematical Association of America, Washington, DC, 1996.

Steven R. Benson, Center for Mathematics Education, Education Development Center, 55 Chapel St, Newton, MA 02458-1060; sbenson@edc.org.

Bradford R. Findell, Department of Mathematics Education, 105 Aderhold Hall, University of Georgia, Athens, GA 30602-7124; bfindell@coe.uga.edu.

On Driving Students to Abstraction

Paul Fjelstad

Abstract. Six first-year college students are recruited to take part in an experimental class that uses neither lectures nor books. The intent is to have them build abstract structures to account for concrete experiences. From flipping simple objects such as a penny or a piece of paper, they come up with some specific groups. From a special deck of 16 cards they come up with something akin to linear operators on a vector space, except the field is replaced by a semiring. Part of the fun of proceeding in this way is that the participants get to invent the notation and the names of things, which keep changing as the systems evolve and new ideas emerge.

1 Introduction

This is the story of an experiment in providing a small group of first-year college students a chance to do some mathematical thinking on their own. The intent was to move from the concrete to the abstract. I knew the start would be with a penny and a piece of paper as concrete objects, but I had no preordained idea of where we would end up, and the end result was indeed a surprise to me.

The experiment was part of the Aleph Program, a project supported by a LOCI (Local Course Improvement) grant from the National Science Foundation. It took place in the 1978–79 school year. The setting was the Paracollege, a small college within St. Olaf College, which offered an alternative route (with tutorials and examinations) to a B.A. degree. It encouraged experiments and in this case six students ventured to participate.

2 Penny

To attract some students to the Aleph Program, a fairly extensive campaign was put into action when students first arrived on campus. The first activity was a contest to see how few pennies were needed to build a structure on a table such that one of the pennies extends completely beyond the table. This raised many interesting questions and, quite unexpectedly, led finally to a definition reminiscent of a well-formed formula in formal logic. The first student to succeed used six pennies, and, as the contest went on, this number was pushed down to four, to two, and then finally to one. (The parameters didn't rule out turning the table on its side.) Seeing how definitions can clearly specify a problem and also provide a surprising result was a good start.

Having played with pennies, it was appropriate at the first session for those who volunteered for the experiment to take a penny and try to get them to come up with a mathematical system to describe some aspect of its behavior. After puzzling over this and trying several ideas that were dead ends, they considered the act of picking the penny up from the table and putting it back on the table, in which case they could either flip it over or not flip it over. This resulted in the table,

$$
\begin{array}{c|cc}
\rhd & f & n \\
\hline
f & n & f \\
n & f & n
\end{array}
$$

where "$f \rhd f = n$" is read as "flip followed by flip is equivalent to no-flip." Given this system, they found other interpretations for $(f, n, \rhd)$ such as (odd, even, add), (negative, positive, multiply), (false, true, iff), and $(1, 0, +$ (mod 2)). These in turn suggested other binary operations on 2 elements (e.g., other logical connectives), which in turn suggested other things in the world to be interpreted (e.g., on-off switches). With this little taste of the ongoing dialectic between abstract mathematical systems and structures describable by them, they forged ahead, finding all 16 binary operations on 2 elements, and reformulating their notation for them again and again as they discovered more and more structure in the whole system. They surprised their tutor by defining operations on these operations and finding simple schemes for generating the 16×16 tables that resulted. They also found all the unary operations, and having done binary and unary, they then wondered about nullary, which turned out to be fruitful, especially for later developments. (Binary, unary, and nullary operations take respectively 2, 1, and 0 arguments.)

3 Paper

To advance beyond 2-element systems, for which $\{0, 1\}$ had become the set of choice, the penny was replaced by an ordinary sheet of paper. In flipping it and rotating it, one student who had taken some flying lessons, blurted out, "pitch, roll, and yaw", which, for each a $180°$ turn, resulted in the following table.

$\rhd$	n	p	r	y
n	n	p	r	y
p	p	n	y	r
r	r	y	n	p
y	y	r	p	n

In terms of a conventional x, y, z-coordinate system with mutually perpendicular axes, p, r, y are $180°$ rotations about the x, y, z axes. Since the students were now very conversant with the 2-element systems, they were challenged to use them somehow to describe this new system. Since $p \rhd r = y$, they let $x = (x_1, x_2)$, with $x_1, x_2 \in \{0, 1\}$, where $x_1 = 1$ if p is done and $x_2 = 1$ if r is done. Thus,

$$n = (0,0), \qquad p = (1,0), \qquad r = (0,1), \qquad y = (1,1).$$

The original penny system, in mod 2 notation, is

$+$	1	0
1	0	1
0	1	0

and by using it, the table for the piece of paper follows by defining the operation $\rhd$ by

$$x \rhd y = (x_1, x_2) \rhd (y_1, y_2) = (x_1 + y_1, x_2 + y_2),$$

which can be expressed equivalently, by defining the ith component, for all i,

$$(x \rhd y)_i = x_i + y_i.$$

In time, the notation (x_1, x_2) was compressed to $x_1 x_2$ (caution—this concatenation of symbols is not multiplication), so the four elements became $\{00, 10, 01, 11\}$, and $\rhd$ also was changed to $+$. Thus, $p \rhd y = r$ was written $10 + 11 = 01$, where componentwise we have $1 + 1 = 0$ and $0 + 1 = 1$.

4 Pod

The last result suggested ways to construct other abstract systems. The operation $+$ on $\{0, 1\}$ was replaced with other binary operations, the elements $x_1 x_2$ were extended to elements $x_1 x_2 \cdots x_n$, and other concrete objects were considered, such as squares, triangles, cubes, and tetrahedrons.

Having now defined many different things, all called mathematical systems, the students were asked to try to generalize and define what a mathematical system might be. This was hard work, but also fun, especially the naming part. If they had been reading a universal algebra book, they would have called what they finally defined an "algebra", but since they weren't, they suggested names like "frame", "husk", "pod", "shell", "ovo", and "novo." It became "pod" when some one suggested that then the elements of the system could be p's, and another then added that if the p's (peas) were thought of as billiard balls, then the things that operate on them could be q's (cues). A binary operation thus became a 2-pronged cue, and for the rest of the semester they persisted in doing "peacueliar" mathematics.

After discussing many different ways to impose structure on a pod, a specific one was given them to play with, namely $P(\Box, {}^\alpha, *)$, where the cues $\Box$, ${}^\alpha$, $*$ were 0-, 1-, and 2-pronged respectively, such that for $p, p', p'' \in P$

$$p * (p' * p'') = (p * p') * p'', \qquad \Box * p = p = p * \Box, \qquad p^\alpha * p = \Box = p * p^\alpha.$$

They did not know that this was ordinarily called a "group" and so it was dubbed a "flock." They struggled the rest of the semester finding examples of flocks, conjecturing what properties flocks might have, and proving and disproving these conjectures. With some coaxing and cajoling they ended up with proofs for Cayley's Theorem, Lagrange's Theorem, and the Fundamental Homomorphism Theorem. In doing this, they would rework proofs again and again, making them shorter and more elegant. When they checked out what they had done at the end of the semester in an algebra book, they had a real appreciation for the refinement process that manifests itself in a textbook, but also for the difficulty of recapturing the motivation for the mathematics when it is presented only in final form.

5   Tracks

Second semester started by searching for analogues of the group theorems developed in the first semester in case more structure was added (e.g., rings) and in case some structure was deleted (e.g., monoids). Things began to bog down at this point. With the usual flow of spontaneity and enthusiasm slowing to a dribble, it seemed prudent to return to something more concrete. There was a need, like Antaeus of Greek mythology, to touch the earth in order to regain strength and vitality for new forays into the world of abstractions. To this end, the students were given the following situation to play with. Consider a set of square cards with two marks on the left edge of each card and two marks on the right edge. Now draw as many straight lines as desired from the marks on the left to the marks on the right. Since there are four possible lines, there are $2^4 = 16$ different ways of doing this.

The students were given the assignment to define some binary operations on these cards. If one thinks of the cards as being transparent, then superimposing one card on top of another produces a result that can be considered a card. For a good reason, to be discovered shortly, this operation ended up being designated by the "$\vee$" symbol. Thus, for example,

The first method of naming the cards involved labeling the four possible lines in the following way.

Then, to use a previous idea, the card is named $X = X_1 X_2 X_3 X_4$ (using concatenation), where $X_i = 1$ if line i is present. The example above is thus

$$1100 \vee 1010 = 1110.$$

This is quite reminiscent of previous work with the piece of paper, only the 2-element system needed here is

$$\begin{array}{c|cc} \vee & 1 & 0 \\ \hline 1 & 1 & 1 \\ 0 & 1 & 0 \end{array}$$

namely, the logical connective "or." The operation $\vee$ on cards was then defined by

$$(X \vee Y)_i = X_i \vee Y_i,$$

which is read "line i is there in $X \vee Y$ just in case it is there in X *or* it is there in Y."

Another binary operation resulted upon setting two cards next to each other.

$$\boxed{\bigtriangledown} \,\, = \,\, \boxed{}$$
$$\;\; X \;\; Y \qquad\quad XY$$

This example shows there are two ways in the combination to go from the upper left mark to the upper right mark, and if either way is present, then line 1 is present in XY. The notation for multiplication (caution—context must now be used to discern multiplication from concatenation) was used because by using multiplication on the set $\{0,1\}$, namely

$$
\begin{array}{c|c|c}
 & 1 & 0 \\
\hline
1 & 1 & 0 \\
\hline
0 & 0 & 0 \\
\end{array}
$$

which is also the logical connective "and," one can say the above in symbols by

$$(XY)_1 = X_1 Y_1 \vee X_2 Y_3$$

which is read "line 1 is in XY just in case line 1 is in X *and* line 1 is in Y *or* line 2 is in X *and* line 3 is in Y."
Similarly,

$$(XY)_2 = X_1 Y_2 \vee X_2 Y_4, \quad (XY)_3 = X_3 Y_1 \vee X_4 Y_3, \quad (XY)_4 = X_3 Y_2 \vee X_4 Y_4.$$

A certain pattern can be discerned here, but it's not an easy one to remember. At this point, the advantage of different students adapting different approaches became clear. Another student experimented with naming the lines on the card by labeling the marks in the following way.

Then $X = {}_1X_1 {}_1X_2 {}_2X_1 {}_2X_2$ (concatenation), where ${}_iX_j$ is 1 if line ij is there. Then the four cases above end up being expressed simply by

$${}_i(XY)_j = {}_iX_1 {}_1Y_j \vee {}_iX_2 {}_2Y_j = \bigvee_k {}_iX_k {}_kY_j,$$

where $\bigvee$ is defined for $\vee$ in the same way that $\sum$ is defined for $+$, and it's understood that k runs through all values in the set $\{1,2\}$.

I was quite astonished to see this result. Although the students were not familiar with this pattern, I recognized it as matrix multiplication, that operation which students, upon first meeting, often find somewhat mysterious and unmotivated. Simply let i and j refer respectively to row and column in a 2×2 matrix. Explicitly,

$$XY = \begin{pmatrix} {}_1X_1 & {}_1X_2 \\ {}_2X_1 & {}_2X_2 \end{pmatrix} \begin{pmatrix} {}_1Y_1 & {}_1Y_2 \\ {}_2Y_1 & {}_2Y_2 \end{pmatrix} = \begin{pmatrix} {}_1X_1 {}_1Y_1 \vee {}_1X_2 {}_2Y_1 & {}_1X_1 {}_1Y_2 \vee {}_1X_2 {}_2Y_2 \\ {}_2X_1 {}_1Y_1 \vee {}_2X_2 {}_2Y_1 & {}_2X_1 {}_1Y_2 \vee {}_2X_2 {}_2Y_2 \end{pmatrix}.$$

Here the students had a concrete model involving lines connecting points with the subscripts telling the story as one moves from left to right. With this mode of thought, the cards were thought of as tracks along which things could move, and $\mathcal{T}$ was the set of 16 tracks. The notation was evolving, and with the ij convention for naming lines, the definition for $\vee$ had to be rewritten as

$${}_i(X \vee Y)_j = {}_iX_j \vee {}_iY_j.$$

Further, it was found that each track could be expressed as a sum over the four tracks having only one line. To symbolize this, $\boxed{ij}$ was the track with only line ij, $\square$ was the track with no lines, and left multiplication of a track by 1 or 0 was defined as follows.

$$0X = \square, \qquad \text{and} \qquad 1X = X.$$

Then for any track X,

$$X = \bigvee_{i,j} {}_iX_j \,\boxed{ij}.$$

One other track given a special symbol is the one with two parallel lines. It was named "I" since it was the identity for multiplication. In particular, ${}_iI_j = 1$, for $i = j$ and 0 otherwise. The identity for $\vee$ of course is $\square$.

The students determined the algebraic properties of the system $\mathcal{T}(\square, \text{I}, \vee, \cdot)$ and found it inherited most of the properties of $\{0,1\}(0,1,\vee,\cdot)$. In my language, the latter is a semiring (additive inverses are not required) and the track system is also a semiring. Already at this stage a tremendous amount of algebraic manipulation was available and the students used it to prove things such as the associative and distributive properties. They further had ideas that cards could be generalized to n marks on each edge and how other mathematical systems, namely other semirings, could be used with essentially the same formalism. This includes, of course, real numbers and complex numbers.

6   Senders and Receivers

By this time some other ideas had been popping up. One of them was that if tracks determine pathways for something to run on, then maybe we could introduce some new objects, called "senders", which would provide the things to run on the tracks. The new objects were half-cards with two marks, labeled "1" and "2", on the right edge. One could then place a dot by as many of the marks as one liked, meaning there are four kinds of senders. Since one can superimpose senders on top of each other and place a sender on the left edge of a track, the development followed closely that just done for tracks. Further, the mirror image of a sender is a receiver, and the notation adopted for receivers tried to do likewise. At the blackboard, a script x was a sender and a backward script x was a receiver. For the printed page, an r will be used for a receiver.

The four senders and receivers are

$$\rceil \quad \rceil \quad \rceil \quad \rceil \qquad\qquad \lceil \quad \lceil \quad \lceil \quad \lceil$$

The special symbols $\overline{1}]$, $\overline{2}]$ and $[\overline{2}$, $[\overline{1}$ were introduced for the objects having just one dot. Then (concatenation)

$$x = x_1 x_2 \qquad\qquad {}_2 r_1 r = r$$

where $x_i = 1$ if dot i is present, and similarly for r. One then has

$$x = \bigvee_i x_i \overline{i}] \qquad\qquad [\overline{i}\, {}_i r \bigvee_i = r$$
$$(xX)_j = \bigvee_k x_{kk} X_j \qquad\qquad {}_j X_{kk} r \bigvee_k = {}_j(Xr)$$

In matrix language, x is a 1×2 matrix and r is a 2×1 matrix. They can thus be multiplied together with xr being either 0 or 1 and rx being a 2×2 matrix. This suggested replacing the dot half-cards with a half-card having two marks on one edge and one mark on the other edge. The multiplications then, which in matrix form are

$$\begin{pmatrix} x_1 & x_2 \end{pmatrix} \begin{pmatrix} {}_1 r \\ {}_2 r \end{pmatrix} = x_{11} r \vee x_{22} r \qquad \text{and} \qquad \begin{pmatrix} {}_1 r \\ {}_2 r \end{pmatrix} \begin{pmatrix} x_1 & x_2 \end{pmatrix} = \begin{pmatrix} {}_1 r x_1 & {}_1 r x_2 \\ {}_2 r x_1 & {}_2 r x_2 \end{pmatrix}$$

would look like this in track form:

and

Thus 0 and 1 can be considered as tracks and the tracks in $\mathcal{T}$ can be generated by senders and receivers. Two special cases of this, which illustrate the delightful surprise that homemade notation sometimes provides, are

$$\overline{i}]\overline{j} = {}_i \text{I}_j \qquad \text{and} \qquad [\overline{i}\,\overline{j}] = \boxed{ij}$$

Two other results akin to these are

$$\overline{j}] X [\overline{i} = {}_i X_j \qquad \text{and} \qquad {}_i X_j \boxed{ij} = [\overline{i}\, {}_i X_j\overline{j}].$$

By splitting $\boxed{ij}$ apart and inserting ${}_i X_j$, one can thus replace the left multiplication of a track by 0 or 1 with track multiplication.

Thus, everything reduces to just tracks, their superimposition and multiplication. The "fundamental particles" of this track world are $\overline{1}]$, $\overline{2}]$, $[\overline{2}$, $[\overline{1}$, which in track form are

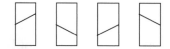

and when so represented, only the first is needed, since the others are simply the result of a pitch, a yaw, and a roll.

For mathematics students this formalism reappears of course in linear transformations on vector spaces, and physics students will later recognize in the *sender-receiver* symbols a form of Dirac's *bra-ket* notation, in which the concept of track is useful, for example, in describing an atom in an atomic beam, where *source-field-detector* = xXr.

7   Concluding Remarks

The students succeeded in using the freedom given to them to come up with mathematical ideas on their own. That freedom, however, was not all that easy to give, for it meant keeping silent at times when I had a great urge to just tell them a theorem and its proof, when the aim, of course, was for them to come up with both. By keeping silent, they would sometimes move off in a direction with a dead end, but recognizing that, and learning how to back out and start afresh, is part of the creative process. On the other hand, sometimes what I presumed to be a dead end, ended up being a fruitful direction in which to go, and I would have squelched that movement by speaking out.

A program like this one does not fit well in a regular college curriculum, but it fit well in the Paracollege, where a prime aim is the fostering of student creativity. The program was supported by NSF for four years, and each year the mathematical ideas that a group of first year students came up with were different, some of which have been reported on earlier [1, 2, 3, 4].

References

[1] P. Fjelstad, *Extending special relativity via the perplex numbers*, Amer. J. Phys. **54** (1986), 416–422.

[2] ——, *Limitless calculus via ethereal numbers*, UMAP Journal **15** (1995), 161–174.

[3] ——, *Extending the Pythagorean Theorem to other geometries*, Math. Mag. **69** (1996), 222–223.

[4] ——, *Mathematics is an Art*, Hum. Math. Netw. J. **21** (1999), 18–26.

Paul Fjelstad, 32991 Dresden Ave., Northfield, MN 55057; `fjelstapstolaf.edu`.

Using Geometry in Teaching Group Theory

Gary Gordon

Abstract. Discrete groups of isometries provide a rich class of groups for detailed study in an undergraduate course in abstract algebra. After developing some geometric tools, we illustrate several important ideas from group theory via symmetry groups. In particular, we consider conjugation, homomorphisms, normal and nonnormal subgroups, generators and relations, quotient groups and direct and semidirect products. Understanding of these topics can be enhanced using symmetries of frieze and wallpaper patterns. We describe several examples in detail and indicate how other geometric investigations could be undertaken. We also indicate how four commercial software packages (*The Geometer's Sketchpad*, *KaleidoTile*, *TesselMania* and *Kali*) can be used to enhance students' understanding.

1 Introduction

The history of group theory is closely tied to measuring the symmetry of some object. Many popular algebra texts now include a section, chapter or collection of chapters that emphasize this approach. See [2], [8] and [16] for typical modern treatments of the subject. I believe the renewed interest in the origins of the subject stems (in part) from a shift in the way we teach our undergraduates—there are more "hands-on" activities and computer based projects than there were five years ago.

As a simple example of how symmetry operations can enhance understanding, consider the following standard homework exercise (4.6 in [2]):

If x, y and xy all have order 2 in a group G, then show $xy = yx$.

This is very easy to prove directly, but a visual approach may be illuminating for students. Consider the Euclidean plane and let x correspond to the operation of a reflection in the x-axis and y correspond to a reflection in the y-axis. Then it is obvious to students that the composition xy corresponds to a $180°$ rotation about the origin. Clearly x, y and xy all have order 2; the conclusion that $xy = yx$ now follows because $180° = -180° \mod 360°$.

Although this approach does not lead to a formal proof of the exercise, it allows students to visualize the exercise concretely. This can mean more to them than the traditional symbol-pushing approach. It also may motivate them to discover other related results on their own. (For example, if every nonidentity element of a group has order 2, then the group is abelian.)

2 Fundamental Properties of Isometries

We begin by reminding the reader of several basic results concerning isometries.

Definition 1. An *isometry* of $\mathbb{R}^n$ is a bijection that preserves distances. More precisely, f is an isometry if, for any pair of points P and Q in $\mathbb{R}^n$, $d(P, Q) = d(f(P), f(Q))$.

It is worthwhile to explain to a class immediately why the set of all isometries forms a group. The group of isometries of the plane is usually called the *Euclidean* group. Also, one should have the students show that the four kinds of isometries of the plane satisfy this definition, i.e., show that translation, rotation, reflection and glide reflection are all isometries.

Working in 1 dimension (instead of 2) can be a beneficial starting point for students. It's easier to visualize the isometries and it allows students to build some confidence and intuition. In particular, it is a straightforward but useful exercise to prove that all isometries $f : \mathbb{R} \to \mathbb{R}$ have the form $f(x) = \pm x + b$, where $b \in \mathbb{R}$.

A more important classification for the remainder of this paper concerns the isometries of the plane. The next result is central.

Proposition 1

    A. *Every isometry of the plane is either a translation, rotation, reflection or glide reflection.*

    B. *Every isometry of the plane is the composition of at most three reflections.*

Proofs of these two results can be found in [6]. Although it is probably not necessary to present or have students develop careful proofs of this proposition in an algebra course, I believe it is *very* important for students to develop geometric intuition by working with isometries. One way to lead students to a discovery of the proposition (with or without a careful proof as a goal) is to use the following exercise.

Exercise 1   (Isometry basics)

1. Given two congruent line segments $\overline{AB}$ and $\overline{A'B'}$, use *The Geometer's Sketchpad* to construct many different isometries, each of which maps $A \mapsto A'$ and $B \mapsto B'$.

2. Show that there are only two distinct isometries that map $A \mapsto A'$ and $B \mapsto B'$.

3. Show that each of these two distinct isometries can be written as a product of at most three reflections.

4. Now if f is any isometry of the plane, show that f is the product of at most three reflections.

5. By investigating the various possibilities (either in *The Geometer's Sketchpad* or by hand), show that the composition of 0, 1, 2 or 3 reflections is always a translation, rotation, reflection or glide reflection.

A few comments on this exercise are in order. Part 2 is believable for students using *The Geometer's Sketchpad*—different students may come up with isometries that appear distinct, but by following the image of a scalene triangle under apparently different maps, they are quickly convinced that there are only two possibilities. Further, part 4 follows from parts 2 and 3 since any isometry f must map the line segment $\overline{AB}$ to the congruent segment $\overline{f(A)f(B)}$, where A and B are any two distinct points in the plane.

It is now possible to ask students if various subsets of isometries form a group. This involves checking the subsets for closure and inverses. As a sample, consider the following example.

Exercise 2   (Group closure). For each of the following sets S of isometries of the plane, determine whether or not S is a group.

1. $S = \{T \mid T \text{ is a translation}\}$.

2. $S = \{R \mid R \text{ is a reflection}\}$.

3. S consists of all translations and all rational rotations, i.e., rotations through angles that are rational multiples of π.

4. S consists of all isometries that fix the origin.

5. S consists of all isometries that fix the x-axis.

6. S consists of all isometries that fix the x-axis or the y-axis.

Problems 4 and 5 are important when we classify the discrete groups that fix a point (the cyclic and dihedral groups) and those that fix a line (the frieze groups). I like these exercises because they use geometry to illustrate group closure. They also show students that composing isometries is important, which is the point of the next section.

3 Three Important Shortcuts

In order to attack serious questions involving groups, it is necessary to understand two important group-theoretic operations as they relate to isometries: composition and conjugation. We begin by fixing notation for each of the four isometries of the plane:

1. Translation by a vector $\bar{v}$: $T_{\bar{v}}$

2. Rotation with center P through an angle θ: $r_{P,\theta}$

3. Reflection through a line l: R_l

4. Glide reflection along the *fixed* vector $\bar{v}$: $g_{\bar{v}}$

We also fix a symbol for each of these four isometries. See Figure 1.

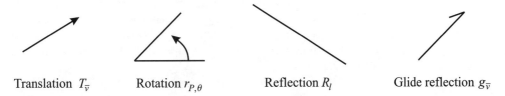

Translation $T_{\bar{v}}$ Rotation $r_{P,\theta}$ Reflection R_l Glide reflection $g_{\bar{v}}$

Figure 1: The four isometries.

Given two isometries, how can we easily determine what isometry corresponds to their composition? We now describe two shortcuts for determining the composition of two isometries, both of which involve the homomorphic images of the isometries. Our first shortcut is the observation that translations and rotations preserve sense, while reflections and glide reflections reverse sense. More precisely, we make the following definition.

Definition 2. An isometry f is *direct* if it preserves right half-planes, i.e., if L is a directed line with right half-plane H, then $f(H)$ is the right half-plane for the line $f(L)$. Otherwise, f is *indirect*.

Direct isometries are sometimes called *sense-preserving*, while indirect isometries are *sense-reversing* or *opposite*. In Figure 2, the left-most drawing represents a right half-plane, the center drawing is the image of this right half-plane under a direct isometry (in this case, a rotation), and the right-most drawing is the image of the same right half-plane under an indirect isometry (in this case, a reflection through a line perpendicular to the directed line).

We now state our first shortcut for composing isometries. The proof follows immediately from the definition.

Proposition 2

1. *The composition of two direct isometries is direct.*

2. *The composition of a direct and an indirect isometry is indirect.*

3. *The composition of two indirect isometries is direct.*

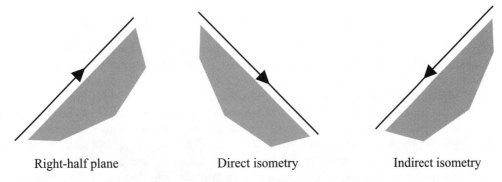

Right-half plane   Direct isometry   Indirect isometry

Figure 2: Direct and indirect isometries.

$$
\begin{array}{c|cc}
* & D & I \\
\hline
D & D & I \\
I & I & D
\end{array}
$$

Table 1: Composing direct and indirect symmetries

This cuts down on the possibilities for the students to check when composing isometries, especially when used in conjunction with our next shortcut. The above proposition is summarized in Table 1, where D stands for a direct isometry and I stands for an indirect one.

Students recognize the similarity between this table and the multiplication table for $\mathbb{Z}_2$. (See Exercise 9 in Chapter 1 of [8].) This suggests that a homomorphism is lurking in the background (which, of course, it is). This can be a starting point for the topic of homomorphisms, or simply an easy example.

Our second shortcut involves recording the effect a given isometry f has on a family of parallel, directed lines. Let $\mathcal{L}_{\overline{u}}$ denote the set of all lines parallel to a given unit vector $\overline{u}$, directed as $\overline{u}$ is. Clearly, this family is carried to another family of parallel, directed lines, say $\mathcal{L}_{\overline{w}}$. Thus, while f itself may not map $\overline{u}$ to $\overline{w}$, this isometry *induces* a map on unit vectors, i.e., f induces a mapping of the unit circle to itself. (This time, the induced map is the image of a homomorphism that sends the full Euclidean group to the isometries that fix the unit circle. The kernel of this map is the normal subgroup of all translations.)

The two most important properties of this shortcut are summarized in the next proposition.

Proposition 3 *Let f be a direct isometry, $\mathcal{L}_{\overline{u}}$ denote a family of parallel, directed lines and $\mathcal{L}_{\overline{w}}$ be the image of $\mathcal{L}_{\overline{u}}$ under f, where $\overline{u}$ and $\overline{w}$ are unit vectors.*

  1. *f is a translation iff $\overline{u} = \overline{w}$.*

  2. *If f is a rotation, then the angle between $\overline{u}$ and $\overline{w}$ is the angle of the rotation.*

We now give an example of how these two shortcuts can be used to determine the composition of two isometries.

Proposition 4

  1. *$T_{\overline{v}}\, r_{P,\theta} = r_{Q,\theta}$ for some point Q.*

  2. *$r_{P,\theta_1}\, r_{Q,\theta_2} = r_{R,\theta_1+\theta_2}$ for some point R, unless $\theta_1 + \theta_2$ is a multiple of 2π, in which case $r_{P,\theta_1}\, r_{Q,\theta_2} = T_{\overline{v}}$ for some vector $\overline{v}$.*

  3. *$R_{l_1} R_{l_2} = r_{P,\theta}$, unless l_1 and l_2 are parallel, in which case $R_{l_1} R_{l_2} = T_{\overline{v}}$.*

Proof.

  1. By shortcut 1, the composition of a translation and a rotation is a direct isometry. But this composition does not fix a family of parallel, directed lines, so by shortcut 2, $T_{\overline{v}} r_{P,\theta} = r_{Q,\theta}$ for some point Q.

  2. By shortcut 1, $r_{P,\theta_1} r_{Q,\theta_2}$ is direct. This composition will leave a family of parallel, directed lines invariant iff $\theta_1 + \theta_2$ is a multiple of 2π. Proposition 3 now gives the result.

  3. Again, shortcut 1 tells us this composition is direct. The result now follows as in part 2. ∎

We point out that the usual proofs of parts 1 and 2 of this proposition are much more involved geometrically (although a geometric proof will determine the point Q in part 1 and the point R in part 2). Part 3 is easy to prove directly. It is fun to demonstrate part 3 with *The Geometer's Sketchpad*. To do this, draw a scalene triangle and two lines. By making the two lines mirrors, one can reflect the triangle in each line (and hide the intermediate reflection if desired). Then the dynamic nature of the program allows one to vary the angle between the two lines and watch the doubly reflected image of the triangle move. Furthermore, the angular speed with which the triangle moves is twice the angular speed of the mirror line. This leads to the conclusion that the angle of rotation is twice the angle between the mirror lines. One can also repeat this experiment when the two lines are parallel.

Our final shortcut involves conjugation of isometries and our symbols from Figure 1. We omit the straightforward but long proof.

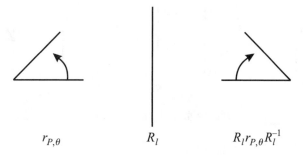

Figure 3: Conjugation of a rotation by a reflection.

Proposition 5 *Let f and g be isometries. Then the symbol for the isometry fgf^{-1} is obtained by applying the isometry f to the symbol corresponding to g.*

As an example of this result, let $f = R_l$ and $g = r_{P,\theta}$, as in Figure 3. Then the conjugate isometry $R_l \, r_{P,\theta} \, R_l^{-1}$ is the rotation $r_{P',-\theta}$, where P' is the image of the point P when reflected through the line l.

Proposition 5 is very useful for determining if a given subgroup is normal. We conclude this section with an entertaining application, due to Tom Brylawski (private communication).

Exercise 3. Let A, B and C be the vertices of a triangle T in the plane. Now perform the following operations on T:

1. Rotate T $180°$ about A; call the resulting triangle T_1, with vertices $A_1(= A), B_1$ and C_1.

2. Rotate T_1 $180°$ about B_1; call the resulting triangle T_2, with vertices $A_2, B_2(= B_1)$ and C_2.

3. Rotate T_2 $180°$ about C_2; continue the labeling process as above.

4. Rotate T_3 $180°$ about A_3.

5. Rotate T_4 $180°$ about B_4.

6. Rotate T_5 $180°$ about C_5.

Show that $T_6 = T$, i.e., the triangle returns to its original position.

To do this exercise, we first let H_P denote the $180°$ rotation (or *half-turn*) about the point P. Then the isometry represented by the first step above is simply H_A. After two steps, the isometry is the composition $H_{B_1} H_A$, using the usual right-to-left ordering on composition. But $H_{B_1} = H_A H_B H_A^{-1}$ by our conjugation shortcut of Proposition 5. Thus $H_{B_1} H_A = (H_A H_B H_A^{-1}) H_A = H_A H_B$. The net effect of conjugation is that *multiplication has been reversed*.

A similar argument shows that, after three steps, the isometry is given by the composition $H_A H_B H_C$. Now note that $H_A H_B H_C = H_D$ for some point D, since this composition is direct (by shortcut 1) and it reverses any family of parallel, directed lines (shortcut 2). Thus, after six steps, the isometry is represented by $(H_D)^2$, which is the identity. This establishes the result.

We remark that other proofs are possible here, but this is probably the quickest. Our proof also makes use of all three of the shortcuts introduced in this section. It is a good exercise to carry out the six steps of the exercise either by hand (with a triangle cut out of paper) or via *The Geometer's Sketchpad*. Students can then guess the answer before seeing a proof.

4 Subgroups of Crystallographic Groups

We now illustrate several topics typically presented in a course that introduces groups.

Example 1 (Generators and relations). Let G be the symmetry group of an infinite strip of V's, i.e., the symmetry group of the pattern

$$\cdots \quad \text{V} \quad \text{V} \quad \text{V} \quad \text{V} \quad \text{V} \quad \text{V} \quad \cdots$$

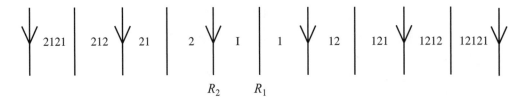

Figure 4: Labeling each strip.

This symmetry group, the infinite dihedral group D_∞, is one of the seven *frieze* groups. See [15] for a detailed presentation of these seven groups. This group is generated by two adjacent vertical reflections, one that is halfway between two adjacent V's and the other that bisects a V. Call these reflections R_1 and R_2, respectively, as in Figure 4. The collection of all lines of reflection divides the plane into a family of vertical, parallel strips. Note also that the collection of all reflections in G fall into two conjugacy classes (by shortcut 3, the conjugation rules). The reflections conjugate to R_1 are precisely the reflections that are halfway between adjacent V's, while the reflections conjugate to R_2 are the reflections through the V's.

Now label the strip between the two parallel lines corresponding to R_1 and R_2 with the identity I, and we call this the *fundamental region*. Then we can label each strip uniquely as follows: Let S be a vertical strip between adjacent reflections and assume inductively that the strips between I and S have already been labeled. Assume S is to the right of I and let S' denote the strip that borders S on the left. (If S is to the left of I, let S' denote the strip bordering S on the right.) Let w be the label for S' and let R' be the reflection corresponding to the border between S' and S. Then label S by $R_i w$, where $i = 1$ or 2 depending on whether R' is conjugate to R_1 or R_2. (This *reversed* multiplication stems from our conjugation rules—see Example 3.) In Figure 4, we write 1 for R_1, 121 for $R_1 R_2 R_1$, and so on.

This gives a labeling of each strip with some word of alternating R_1's and R_2's. This approach, which is dual (in the graph theoretic sense) to the *Cayley graph* of the group, is called the *dihedral kaleidoscope* and is presented for many reflection groups in Coxeter's classic [7].

We now use this example to illustrate several topics. The same analysis can be used on more complicated symmetry groups (e.g., the wallpaper groups). The payoff for working through harder examples is a deeper understanding of the topics involved. We concentrate on our simple example here to illustrate the main ideas; if some students or a class becomes especially interested in this approach, then it might be worthwhile to explore deeper examples.

Topic 1 (Generators and relations). The correspondence between the regions of Figure 4 and the words of D_∞ shows that the reflections R_1 and R_2 generate D_∞. The relations $R_1^2 = R_2^2 = I$ follow from the simple fact that when one crosses the line from region I to region R_1, say, and then crosses the line back again to I, one's path corresponds to the word R_1^2, but the region occupied is I. This makes use of the fact that our region labeling scheme could easily be extended to *paths* from region I to any other region. Thus two paths that begin in the same region and end in the same region must correspond to the same word in G. The "straight-line" or "light-ray" path will give a shortest or reduced word for the region. Longer paths that wind back and forth will give words that can be reduced via the relations to the shortest word.

Topic 2 (Normal subgroups, cosets, quotient groups). There are several ways to find normal subgroups of D_∞. By our conjugation rules, the set of all translations in D_∞ is a normal subgroup. (Note that conjugating a translation T by either of the generators R_1 or R_2 yields T^{-1}. Thus the set of translations is invariant under conjugation.)

This gives two cosets in D_∞, $\mathcal{T}$ and $R_1 \mathcal{T}$, where $\mathcal{T}$ is the normal subgroup of translations. The coset $\mathcal{T}$ corresponds to all regions with labels using an even number of reflections (as these are precisely the translations), while the coset $R_1 \mathcal{T}$ corresponds to all regions whose labels use an odd number of reflections.

Since there are only two cosets, the quotient group is $\mathbb{Z}_2$. To see this via the generators and relations, note that modding out by all translations forces $R_1 R_2 = I$ in the quotient group. No additional new relations are forced, since $R_1 R_2$ generates $\mathcal{T}$. Thus,

$$D_\infty/\mathcal{T} = \langle R_1, R_2 \mid R_1^2 = R_2^2 = R_1 R_2 = I \rangle.$$

This easily reduces to $\langle R \mid R^2 = I \rangle$.

Other subgroups of translations are also normal subgroups of D_∞. For example, if H is the subgroup generated by T^5, where $T = R_1 R_2$ is the shortest translation, then H is normal (by our conjugation rules) and D_∞/H has presentation

$$D_\infty/H = \langle R_1, R_2 \mid R_1^2 = R_2^2 = (R_1 R_2)^5 = I \rangle.$$

This is a standard presentation for the dihedral group D_5. This procedure is entirely analogous to the usual way quotient groups are used to obtain $\mathbb{Z}_5$ from $\mathbb{Z}$. It is an entertaining exercise to color the regions of Figure 4 with ten colors corresponding to the ten cosets generated.

To see how a nonnormal subgroup can be generated, let H be the subgroup generated by $R_1 R_2 R_1$ and $R_2 R_1 R_2$. This subgroup will include every *third* reflection (as well as translations of the form T^{3k}, where $T = R_1 R_2$ is the shortest translation as before). Note that $H \cong D_\infty$ since H is generated by two parallel reflections. Then H is not normal—R_1 is clearly conjugate to $R_2 R_1 R_2$, but $R_1 \notin H$.

In spite of this nonnormality, we can still attempt the *coset coloring* mentioned above. Then we get the left cosets $H, R_1 H$ and $R_2 H$. This induces the left coset coloring of Figure 5, where each region in a given coset is shaded the same.

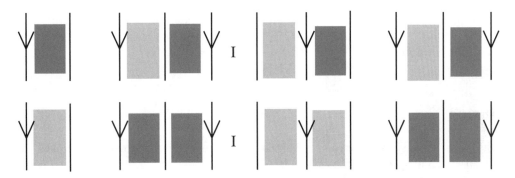

Figure 5: Row one—Left coset coloring; Row two—Right coset coloring

In the same way, we can shade the regions using the right cosets H, HR_1 and HR_2. These cosets do not coincide with the left cosets because H is not normal. This nonnormality is reflected in the coloring—the right coset coloring does not match the left coset coloring.

We conclude by considering direct and semidirect products of groups. Our geometric examples for these topics again involve the frieze groups.

Example 2. Let G be the symmetry group for the pattern

$$\cdots \quad \text{H} \quad \text{H} \quad \text{H} \quad \text{H} \quad \text{H} \quad \text{H} \quad \cdots$$

Then G is generated by two vertical reflections R_1 and R_2, as before, together with one horizontal reflection R_h. The collection of all vertical lines of reflection with the horizontal line partitions the plane into regions, as in Figure 4. Our labeling scheme again gives shortest words for each region and the fact that the line corresponding to R_h is orthogonal to both of the lines for R_1 and R_2 means that R_h commutes with both R_1 and R_2. Thus, we get the presentation

$$G = \langle R_1, R_2, R_h \mid R_1^2 = R_2^2 = R_h^2 =_h R_1 R_h R_1 = R_h R_2 R_h R_2 = I \rangle.$$

Topic 3 (Direct product). The presentation for the group G from Example 2 shows R_h commutes with the other generators. Thus G is a direct product of the normal subgroup generated by R_1 and R_2 (which is isomorphic to D_∞) and the normal subgroup generated by R_h (which is isomorphic to the dihedral group generated by a single reflection, D_1). Thus

$$G \cong D_\infty \times D_1.$$

We conclude this section with another frieze group that has a semidirect product decomposition.

Example 3. This time, let G be the symmetry group for the following pattern:

$$\cdots \quad \cup \quad \cap \quad \cup \quad \cap \quad \cup \quad \cap \quad \cdots$$

Then G contains vertical reflections (through the union and intersection symbols), half-turns (halfway in between the symbols) and horizontal glide reflections (and, of course, translations). If R is a reflection and H is a half-turn, then RH is a glide reflection. (To see this, note RH is indirect by shortcut 1, so RH is either a reflection or a glide reflection. If $RH = R'$ for some reflection R', then $RR' = H$. But RR' is a translation since R and R' are parallel reflections. Thus $RH = g$ for some glide reflection g.) If R is a vertical reflection and H a half-turn whose center is as close as possible to the line corresponding to R, we get the following presentation for G:

$$G = \langle R, H \mid R^2 = H^2 = I \rangle.$$

Thus $G \cong D_\infty$. (In fact the seven frieze groups fall into only four isomorphism classes: $C_\infty, C_\infty \times D_1, D_\infty, D_\infty \times D_1$.)

Topic 4 (Semidirect products). Let G be the symmetry group in Example 3 and let N be the normal subgroup of G generated by all vertical reflections. Then $N \cong D_\infty$. Let R_1 and R_2 be adjacent vertical reflections and H the half-turn between these reflections. Then $HR_1H = R_2$, i.e., conjugation by H interchanges the generators of N. Then $G/N = \langle H \mid H^2 = I \rangle \cong D_1$. Let $M = \{I, H\}$ be the (nonnormal) subgroup generated by H. This gives the semidirect product decomposition

$$G \cong N \rtimes M.$$

5   Conclusion

Almost all of the topics typically encountered in an introductory group theory course can be illustrated with symmetry groups. This is an engaging approach for many students because they see a connection between two areas (group theory and geometry) and get to try out their group theory ideas in a different setting. In addition, the patterns we study are taken from different cultures and times. For example, some of the most beautiful patterns are found in the Alhambra, the fourteenth century Moorish palace in Granada. It was after his trips to this palace in the 1920s and 1930s that Dutch artist M. C. Escher made a detailed study of the possible repeating patterns and developed his own classification scheme for them. See [18] for more information on Escher.

Many computer programs are available to illustrate these ideas. *The Geometer's Sketchpad* [13] is very easy to use and allows the user to define any of the standard isometries (translation, reflection and rotation). *TesselMania* [14] allows one to create one's own repeating patterns, and *Kali* [9] allows one to select an isometry group and then draw a picture that has the selected group as its symmetry group. *KaleidoTile* [10] allows visualization of a symmetry group generated by reflections, including hyperbolic groups and the finite symmetry groups in three dimensions. Links to each of these pieces of software are available from my page at this volume's web site. (See the appendix for details.)

Some suggestions for specific projects appear in [11], while [19] includes several computer-based projects for an algebra class. The bibliography also includes many interesting resources not explicitly referred to in the body of this paper.

References

[1] C. Alexander, I. Giblin and D. Newton, *Symmetry groups of fractals*, The Math. Intelligencer, **14** (Spring 1992) 32–38.

[2] M.A. Armstrong, *Groups and Symmetry*, Undergraduate Texts in Mathematics, Springer-Verlag, New York, 1988.

[3] T. Brylawski, *Notes on Mathematics and Art*, unpublished.

[4] J.H. Conway, *The orbifold notation for surface groups*, in Groups, Combinatorics and Geometry, (Liebeck and Saxl, ed.) London Mathematical Society Lecture Note Series 165, Cambridge, 1992.

[5] H.S.M. Coxeter, *A simpler introduction to colored symmetry*, International Journal of Quantum Chemistry, **31** (1987) 455–461.

[6] H.S.M. Coxeter, *Introduction to Geometry*, 2nd ed., John Wiley & Sons, Inc., 1961, 1969.

[7] H.S.M. Coxeter, *Regular Polytopes*, 3rd ed., Dover, 1973.

[8] J. Gallian, *Contemporary Abstract Algebra*, 4th ed., Houghton Mifflin, 1998.

[9] The Geometry Center, *Kali*, `http://www.geom.umn.edu/software/download/kali.html`.

[10] The Geometry Center, *KaleidoTile*, `http://www.geom.umn.edu/software/download-/KaleidoTile.html`.

[11] G. Gordon, *Using wallpaper groups to motivate group theory*, PRIMUS **VI** (1996) 355–365.

[12] B. Grünbaum and G.C. Shephard, *Tilings and Patterns*, W.H. Freeman, 1989.

[13] Key Curriculum Press, *The Geometer's Sketchpad*, Emeryville, CA, `http://www.keypress.com/`.

[14] The Learning Company, *TesselMania*, Novato, CA, `http://www.learningcompanyschool.com`.

[15] G.E. Martin, *Transformation Geometry*, Undergraduate Texts in Mathematics, Springer-Verlag, New York, 1982.

[16] V.V. Nikulin and I.R. Shafarevich, *Geometries and Groups*, Springer-Verlag, Berlin, 1987.

[17] D. Schattschneider, *The plane symmetry groups; their recognition and notation*, Amer. Math. Monthly, **85** (1978) 439–450.

[18] D. Schattschneider, *Visions of Symmetry*, W. H. Freeman & Co., 1990.

[19] D. Schattschneider, *Visualization of group theory concepts with dynamic geometry software*, in Geometry Turned On, Notes 41, MAA, 1997.

[20] D. Seymour and J. Britton, *Introduction to Tessellations*, Dale Seymour Pub., 1989.

[21] D.K. Washburn and D.W. Crowe, *Symmetries of Culture*, Univ. of Washington Press, Seattle, 1988.

Gary Gordon, Department of Mathematics, Lafayette College, Easton, PA 18042; gordong@lafayette.edu.

An Abstract Algebra Research Project: How many solutions does $x^2 + 1 = 0$ have?

Suzanne Dorée

Abstract. This paper describes the implementation of a student research project in an abstract algebra class and includes the rationale for incorporating a research project, the process of developing a successful research question, guidelines for implementation in the classroom, and a description of outcomes. The project centered on the title question and provided a concrete setting for investigating the connections between polynomial equations, matrix rings, modular integers, unique factorization, and roots of -1.

1 Introduction

Thinking about my upcoming abstract algebra course, I asked myself the following questions:

> How can I help students better understand abstract algebra? How can I introduce undergraduate students to mathematical research?

By incorporating a research project into my abstract algebra course, I was able to answer both questions at once. Having made that decision, my next challenge was to design a project and then fit it into the rest of the course. In this paper I describe what I did and how it worked. To whet the reader's interest, I also reveal some of the mathematics involved, but for a more rigorous discussion see [1].

2 Why Incorporate a Research Project?

I was frustrated that some of my abstract algebra students found the course to be difficult, understood the theory only superficially, and were not very interested in the concepts. My reasons for incorporating a research project were to increase my students' understanding of the course concepts, to give concrete applications of the theory, and, hopefully, to have some fun.

I also wanted students to learn more about mathematical research and experience it themselves. Although few of my students were continuing to graduate school in mathematics, I believed that research skills such as posing questions, checking special cases, and recognizing patterns would be transferable to other careers. I also believed that mathematics majors ought to have an idea of what mathematicians do.

While there are some undergraduates who publish work comparable to a professor's, I was interested in making the research experience accessible to my students even if they were not at that level. I extended my definition of undergraduate research to include student work on open-ended problems. I still expected that the student participate in framing the problem and that the results be both original to the student and not specifically known by the project advisor. Such projects are also referred to as "investigations" or "explorations" but I used the term "research" to help students see the connection to more sophisticated levels of research.

3 Finding a Good Research Question

After deciding to incorporate a research project into my abstract algebra course, my next task was to find a topic. I thought carefully about what had been interesting in previous years. Students seemed to like polynomials and solving equations, and I enjoyed this area as well. The theory of equations was accessible since the course I teach covers rings before groups [2].

I chose the square root of -1 as the specific topic. Former students had expressed interest in whether it really existed and, if so, how it was rigorously defined. This topic had some historical connections to algebra and was important in its own right. Since my doctoral research had been in character theory, I was confident that I could answer their questions.

I decided to use a question as a starting point. I considered using "How can we define the square root of -1?" but decided there was too much of the answer in the question that might have limited further investigation. Next, I posed "Can $x^2 = -1$?", which I quickly improved to "Does $x^2 + 1 = 0$ have a solution?" However, I was discouraged by the philosophical tone of the question and wanted something more computational. Ultimately, I used the question "How many solutions does $x^2 + 1 = 0$ have?" Another equation that would probably yield similar success is the defining equation of the golden ratio: $x^2 = x + 1$. Alternatively, other roots of -1 could be explored using the equation $x^n + 1 = 0$.

In retrospect, the question I used worked well because it met several criteria, listed below.

- There are few prerequisites to understanding the question.

- The question has some easy, accessible answers but it also has more sophisticated answers that build on course content.

- Students can experiment using a computer.

- The topic has historical importance.

- The topic has relevance throughout the course.

- I know something about the subject.

4 Implementation

There were eight students in my abstract algebra class, six of whom were mathematics majors. Most of them were accustomed to working on projects during class in small groups.

The first day of class I began by posing the question "How many solutions does $x^2 + 1 = 0$ have?" After writing for several minutes, the students compared answers in small groups. It was important that they had time to think individually first to prevent "group think."

As expected, different students had different answers. Some said "zero" and others said "two"; I helped them rephrase their answer to "no solutions in $\mathbb{R}$ but two solutions in $\mathbb{C}$." Most students recognized that the answer depended on the context, which was, of course, one point of the exercise.

Surprisingly, some students thought there was only one solution: i. Others convinced them that $-i$ was also a solution. From this mistake, students began to think about what would later be their first theorem: The additive inverse of a solution is also a solution.

After the group work, I explained that we could discuss the question in any context in which addition, squaring, 1, and 0 made sense. Students contributed examples of places they had seen these operations and identities, including the usual number rings, matrices, functions, and sets.

Next, each student generated a list of questions for possible future investigation, such as what is the meaning of i, how to find the solution(s) of a polynomial equation, and whether there were rings with other numbers of solutions. These questions helped frame future exploration and connect the project to the course content.

We returned to the project six days throughout the semester, and each time we spent most of the hour working on a specific question I posed. In-class work was followed by an assignment to continue their exploration and write up their progress at home. The specific questions I assigned in class modeled the research process: define the question, look at special cases, pose further questions, relate conjectures to known theory, and prove conjectures. The six questions I used were as follows.

- How many solutions does $x^2 + 1 = 0$ have in $\mathbb{Z}_{12}$?

- How many solutions does $x^2 + 1 = 0$ have in the ring of 2×2 matrices over $\mathbb{R}$?

- How can you prove your current conjectures?

- For which integers n does $x^2 + 1 = 0$ have a solution in $\mathbb{Z}_n$?

- How do your results connect to the theory of polynomials, especially the idea of uniqueness of factorization?

- How can you prove your recent conjectures?

We first revisited the project several weeks into the course, after reviewing properties of the integers and the integers mod n. They looked at $\mathbb{Z}_n$ for various values of n, which is easily done by hand for small values of n and provides opportunity for practicing modular arithmetic.

Many of my students noticed one pattern right away: If $n = k^2 + 1$ for some integer k, then k and $-k$ are solutions. I mention this result for two reasons. First, I did not explicitly think of it myself, which made their discovery of it fun for me. Second, it concretely foreshadows the quotient field construction of roots of a polynomial.

There are other values of n for which $\mathbb{Z}_n$ has solutions to $x^2 + 1 = 0$. For large values of n, the calculations are best done using a computer or programmable calculator. It turns out that the prime factorization of n determines whether $\mathbb{Z}_n$ will have square roots of -1 and, if so, how many there will be. For a statement of these results and references to proofs, see [1].

Another fruitful discussion occurred when I asked students to consider the matrix analog of the equation, $X^2 + I = 0$, where X is a 2×2 matrix. They were able to solve quickly for all (infinitely many) solutions with real-valued entries, but struggled when I asked them to determine which solutions had integer-valued entries and whether there were additional solutions with complex-valued entries. Some students delved into connections to the quaternions, to matrix rings over $\mathbb{Z}_n$, or to larger dimensional matrices in their individual research.

For two of the days devoted to the project, students worked on proving their conjectures with feedback from their peers as well as myself. Some students were only able to answer the questions I had posed, but others quickly progressed to attempting to prove their own conjectures.

After the first conjecture-proving day, they wrote a preliminary report. This assignment encouraged students to begin writing up their work before the end of the term and gave me an opportunity to give them written feedback.

At the end of the term students wrote a final report, which comprised 25% of their course grade. Half of the report grade was based on their calculation and proof of the correct number of solutions in $\mathbb{R}$, $\mathbb{C}$, several values of $\mathbb{Z}_n$, and 2×2 matrices over $\mathbb{R}$; for a high grade, I expected students to have additional substantial conjectures and results. Another quarter of the grade was determined by their competence in relating the question to the ring theory and theory of polynomials; I gave extra credit for connections to groups. The last quarter of the grade I reserved for writing, i.e., clear explanations of the mathematics and proper use of mathematical terminology and notation. To maximize the time students spent on the mathematics, I provided a rough outline for the paper. Although I made some deductions for poor appearance, spelling, or grammar, I tried to keep the focus on assessing their understanding and exposition of the mathematics.

5 Outcomes

Incorporating a research project took time away from covering material and required that I spend more time grading papers and working with students outside of class. I think, however, the project was well worth these costs. Covering less material gave students the leisure to learn the covered topics more thoroughly. Spending more time working with students outside of class gave me the opportunity to work one-on-one with students and get to know my students better. I cannot say that spending more time grading was enjoyable, but it did increase my understanding of my students' mathematical challenges and talents.

I was impressed with the learning outcomes of the project. My students understood abstract algebra at a deeper level and raised questions that were answered later in the course, which in turn increased their interest in the course. They learned more about mathematical research and gained experience doing it. One of my students became so interested in research that she continued working on the project with me during the following semester. Like many other mathematics professors, I am working on getting undergraduates interested in research and developing the required skills. My experience shows that an in-class project is an excellent starting point!

References

[1] S. Dorée, *How Many Solutions Does $x^2 + 1 = 0$ Have? An Abstract Algebra Project*, PRIMUS **10** (2000), 111–122.

[2] T. Hungerford, *Abstract Algebra: An Introduction*, second edition, Saunders College Publishing, New York, 1997.

Suzanne I. Dorée, Mathematics Department, Augsburg College, 2211 Riverside Avenue South, Minneapolis, MN 55454; doree@augsburg.edu; http://www.augsburg.edu/math/.

Part II

Using Software to Approach Abstract Algebra

Laboratory Experiences in Group Theory: A Discovery Approach

Ellen J. Maycock

Abstract. Traditionally, group theory is taught using the "theorem-proof-example" format. Although this method of presentation is very satisfying to mathematicians, many students have difficulty learning the concepts with this approach. In this paper, I will describe how my use of the software package *Exploring Small Groups* transformed my abstract algebra classroom. The materials I developed, which reflect my desire to have the classroom be a creative learning environment for my students, have been published by the Mathematical Association of America as the volume *Laboratory Experiences in Group Theory*.

1 A Group Theory Class with Technology

I began using the software package *Exploring Small Groups* (*ESG*) for demonstration purposes in the spring of 1990. One day, instead of answering a question asked by a student in class, I sent the students to the computer lab to generate some conjectures. The following day, they were asked to prove or disprove several of the conjectures made by their classmates. One student commented: "How can we work on these when we don't know whether they are true?" At that point, I realized that my students had had little opportunity to experience mathematical discovery.

When I taught the class again, the course was organized around weekly laboratories that developed the basic ideas of group theory through examples and discovery. The class met four hours per week: two hour-long lecture/discussion sessions held in a regular classroom and a two-hour session held in a computer laboratory. The students worked in pairs in the lab, and recorded their data during the lab session. Each week, a new topic was examined in the lab. I quickly learned that the students needed to work out at least one example by hand before using the computers. This guaranteed that the students did not view the computer as a "black box," generating mysterious answers. After a basic calculation was done once or twice, however, it was effective to have the computer complete the computations. Laboratory reports were required for each lab. Throughout the semester the students assembled a great deal of information about a basic collection of eight to ten groups. Follow-up classroom sessions emphasized the theory and proofs that are so fundamental to this course. The laboratory materials, which I discuss in more detail below, have been published by the Mathematical Association of America as *Laboratory Experiences in Group Theory* [3]. These labs should be considered supplementary to a standard introductory textbook chosen for the course.

I have always believed that true learning takes place when students write up their own lab reports. The fact that the questions on the labs did not have quick, standard answers was a new experience for many students. Primarily, the lab reports are a place where students can summarize the patterns in their data. Usually, I ask a series of questions to step them through the lab itself, and the report must contain the answers to these questions. At the end of the reports, the students are asked to make conjectures. Conjecture-writing is an excellent way for students to play with the concepts of group theory. In some sense, it doesn't matter what the statement of the conjecture actually is. To write a conjecture, a student needs some kind of understanding about the nature of the mathematical object that is being investigated. In an introductory algebra course, students are primarily learning the basic structures rather than a long list of theorems. Their conjectures sometimes anticipated material in the text, often with different phrasing and notation. It was especially nice when a conjecture gave me the basis for a new classroom discussion. I have never worried about students "cheating" by looking in their textbook for an answer; I would love for my students to read their textbook!

2 Using *Exploring Small Groups* in a Classroom Setting

Exploring Small Groups (*ESG*) [2] was written by Ladnor Geissinger of the University of North Carolina as a teaching tool for group theory. He has shown his deep understanding of how students learn by designing a simple-to-use program that emphasizes the basic constructions of group theory. The software has a library of all the groups of order 16 and below; this is surprisingly adequate for an introductory course. My students quickly learn the patterns of the Cayley tables of many of the groups of small order, and they base subsequent conjectures on their recognition of these patterns. The five groups of order 12 illustrate most of the important distinctions we wish to make at this level, and the groups of order 16 provide much material for experimentation. The software has menu choices that allow students to construct subgroups, quotient groups, and endomorphisms. In addition, students may randomly generate Cayley tables and then check to see if the group axioms are satisfied. The use of color is especially inspired. I have been able to use the program successfully with color-blind students if they are paired with students who can see all the colors on the screen. The simplicity of *ESG* has allowed me to use technology quickly and easily in my algebra classroom.

It is important to introduce students slowly to both the laboratory approach and the software. In the first lab, students work with cardboard or paper models of a triangle, rectangle and square. As students begin to use *ESG*, they need to understand that a variety of notations can illustrate the same mathematical structure. Therefore, I have been careful to show how their textbook and the program can be correlated, using the feature of *ESG* that renames the elements of the group. In the next few labs, cyclic groups and subgroups are discussed and the students are shown how to construct subgroup lattices. Students must initially use a trial-and-error approach to create subgroups, constructing all possible subgroups with generators added one by one. I only smile—and do not answer—when students ask: "How do we know when we have all the subgroups of a group?" Students usually come up with Lagrange's Theorem as a conjecture on their own; I'm careful to include A_4 as an example so that they don't assume that the converse of Lagrange's Theorem is correct. Specific menu choices in *ESG* make it easy for students to compute the center and commutator subgroups of a group. After completing the labs for these special subgroups, students have a sizable collection of annotated subgroup lattices with which we work for the rest of the semester.

The true strength of *ESG* lies in how it handles normal subgroups. After generating a subgroup, one may choose to look at the Cayley table with the elements reordered and colored by left cosets. It is easy to discuss whether the coset operation is well-defined from this configuration. If the subgroup is normal, the program transforms the blocks of colored elements in the body of the Cayley table into solid blocks of colors. That the factor group is constructed from the elements of the group, but is not a subgroup of the group, has always been the most difficult concept for me to communicate to my students in introductory group theory. This visual presentation makes the factor group easily understood by all my students. By this time in the course, students are often able to identify the pattern presented by the color blocks of the factor group as one of the familiar groups we have already studied.

Later in the course, several labs show students how to construct endomorphisms. The students are initially quite frustrated, as their efforts seem to be primarily trial-and-error. This is a topic for which paper-and-pencil work is imperative. Once an endomorphism is constructed by *ESG*, it is easy to identify the image and kernel of the mapping on the computer screen. One of the best lab experiences I've had with my classes came one year toward the end of the semester. A hint in the textbook led us to ask the question:

> Given a normal subgroup N of a group G, does there exist an endomorphism T of G with $\ker(T) = N$? If the answer to the question is no, what is the obstruction to this being true?

Students were asked to generate a table with the following information: the group G, all normal subgroups of G, and, for each normal subgroup N of G, the rule of one endomorphism whose kernel is N, the image of this endomorphism, and the factor group G/N. For the groups we considered, two had one normal subgroup for which no endomorphism could be constructed. Some students were able to articulate the correct conclusion immediately; others needed to be asked some leading questions to realize that those two groups did not have any subgroups isomorphic to G/N. I have learned in all my laboratory courses that a lab is only as good as its follow-up in class, and this lab had a great one: the Fundamental Homomorphism Theorem.

In subsequent years, I have varied my use of the laboratory. Weekly lab sessions often seem too frequent, and I currently use 6–8 labs over the course of the semester. Sometimes, I assign portions of the labs as homework assignments rather than devote class time to the activity. I like splitting the work among the students in the class and having students share their data with the class. This allows for even more examples to be generated, and the class as a whole can construct conjectures. The most important addition has been to incorporate small research projects at the end of the semester. After learning the basic concepts, my students have been able to investigate open-ended questions with either *ESG* or the more sophisticated software package *GAP* [1]. Students were excited about the possibilities for research once they had the power of technology available to them, and several have been able to prove their own conjectures after generating convincing data.

3 The Roles of Technology in the Upper-level Classroom

Technology can play several different roles in the teaching of abstract mathematics. Most importantly, software can illustrate difficult concepts. We have all seen how useful technology can be in creating visual images for our students. In addition, technology can bridge the gap between intuitive understanding and the formalism that must be mastered in an upper-level course. This can be especially useful, for instance, in a real analysis course, where students struggle to see how the formalism connects with the intuitive ideas they previously learned in calculus. Students develop intuitive ideas about limits and continuity in calculus, then often wait several years to see the formal proofs of analysis. For material that has been learned well at an elementary level, technology can help students explore more advanced material creatively and independently. I have seen this happen repeatedly in an upper-level geometry course, where students are able to build creatively on their background from high school.

None of these explanations answers the question: *Why does discovery seem so effective in elementary group theory?* I believe it is because the subject matter is new to virtually all students. The underlying goal of an introductory abstract algebra course is to introduce the basic concepts or constructions. Students develop an intuitive sense of the concepts as they investigate examples by hand and with software. They then formalize the concepts into the standard language of mathematics. The interplay of intuition and formalism occurs with each laboratory experience. The conceptual understanding and the process of formalizing these concepts are done simultaneously in an introductory abstract algebra course, and technology is an aid in this process.

In all my courses, I am concerned about trying to establish an environment where students can have a creative mathematical experience. In my view, the greatest value of the laboratory experience is that students actually experience those frustrations that ultimately lead to the satisfying "a-ha" moment. The instructor needs to be aware that not all students respond well to the discovery approach to learning. However, my students have been much more engaged in this very abstract material than when I taught the course using more traditional methods, and the classes where I have used the laboratory have been exceptionally lively. One student comment summarizes the important impact of the laboratory:

> I myself am truly grateful for the laboratory component. Work on the computer helped to make the abstract theory more concrete. One of the best things about the labs was that we formed our own conjectures about the patterns we saw. I believe that the progression of (1) lab, (2) conjecture, (3) class discussion, and (4) proof was highly beneficial in gaining understanding of the abstract material of the course.

References

[1] The GAP Group, *GAP — Groups, Algorithms, and Programming*, Version 4.2; Aachen, St Andrews (http://www-gap.dcs.st-and.ac.uk/gap).

[2] L. Geissinger, *Exploring Small Groups: A Tool for Learning Abstract Algebra*, now only available bundled with [3].

[3] E. Parker, *Laboratory Experiences in Group Theory: A Manual to be used with Exploring Small Groups*, Mathematical Association of America, Washington, DC, 1996.

Ellen Maycock, Department of Mathematics, DePauw University, Greencastle, IN 46135; emaycock@depauw.edu; http://www.depauw.edu/acad/mathematics/emaycock.htm.

Learning Beginning Group Theory with *Finite Group Behavior*

Edward Keppelmann, with Bayard Webb

Abstract. In 1989, Ladnor Geissinger (University of North Carolina at Chapel Hill) developed an extraordinary DOS program called *Exploring Small Groups* (*ESG*) for students learning beginning group theory. In 1996, Ellen Maycock Parker from DePauw University augmented *ESG* with a series of labs and instructional materials. In this paper we describe a *Windows*-based program called *Finite Group Behavior* (*FGB*) that improves on the capabilities and instructional power of *ESG*. The program is available free of charge from our web site. In this article we discuss the features and philosophy of *FGB* and we offer some suggestions for its use. In conjunction with Parker's labs, our perspective gives the instructor of beginning abstract algebra some good ideas about using *FGB* to convey the intricacies of group theory as well as the sense of discovery that any research mathematician knows intimately.

1 Introduction

As an undergraduate during the fall of 1994, at the University of Nevada Reno, Bayard Webb was a student of Ed Keppelmann in Math 331: Groups, Rings, and Fields. For the portion of the course devoted to group theory, the main concepts were motivated by a computer component based on the DOS software *Exploring Small Groups* (*ESG*), by Ladnor Geissinger [4] and labs written by Ellen Maycock Parker [8]. With a blend of interests as a highly motivated and curious undergraduate mathematics major and a talented computer programmer, Webb took it upon himself to rewrite and expand on the capabilities of *ESG* in a *Windows*-based environment.

Unlike advanced computational tools such as *MAGMA* [6] or *GAP* [3], *Finite Group Behavior* (*FGB*) [5] is intended as a teaching tool for those at the beginning level of group theory. One of the first steps in the study of groups can be the study of concrete examples, in the form of their Cayley tables, along with just enough functionality so that subgroups, cosets, and homomorphisms can be easily constructed. While a typical course should not be confined to studying only the finite groups available in *FGB*, this is an important starting point so that a broad conceptual understanding can pave the way for other investigations. Too much computational power can make answers appear in a black-box fashion, thereby short-circuiting the learning process. In contrast, what *FGB* does can be replicated on a smaller scale (and often should be) by the beginning student with pencil and paper.

The outgrowth of Geissinger's original conception of *ESG* as a learning tool, along with Webb's persistence and programming talent, is now available from our web site [5]. (Downloading details can also be found in the appendix and on the web site for this volume.) After a brief registration questionnaire, anyone is welcome to use the software free of charge. As well as keeping future users informed about upgrades, the registration information is used as evidence of need and interest to request support in the form of grant applications, faculty release time, and other forms of recognition for future developments. Suggestions, criticisms, and ideas for improvements are always encouraged.

2 Learning objectives

Abstract algebra is arguably one of the most difficult courses in an undergraduate mathematics curriculum. Traditionally, the abstract nature of the course, coupled with a lack of accessible examples, creates a situation in which few students are able to grasp and appreciate the subject at a suitable level. As in a beginning linear algebra course (but perhaps even more difficult here), it is important to teach the abstract nature of the course while still finding the time to suitably motivate the material with meaningful examples. Without a variety of such examples, the student can easily be left feeling that he or she has just acquired a good bit of mostly useless abstract nonsense.

Because the software is really just a computational tool, and not a collection of pre-organized lessons, it possesses a vast variety of potential uses. We sincerely hope that our work leads to new and improved lab manuals such as [4]. In this paper we mostly content ourselves with a discussion of the power of the software and an overview of its basic uses. However, in order to illustrate the possibilities, we have also included a presentation of some rudimentary examples for learning by using the program. We are always eager to consider adding new functionality to *FGB* by adding features that the academic community thinks are necessary. We will close the paper with some of the additions we are currently considering.

In specific terms, we believe *FGB* is well-suited for teaching the following concepts and objectives:

- The axioms of a group and how these are verified from a Cayley table for the case of a finite group.

- Basic notions in a group such as subgroups and orders of elements.

- Group isomorphisms (especially how to recognize when they exist and how to construct them).

- Cosets, normal subgroups, and factor groups (often the hardest concepts for even the best students).

- Homomorphisms and the fundamental isomorphism theorem (i.e., for $\varphi : G \to H$ we have $G/\ker\varphi \cong \varphi(G)$).

- Learning to reason efficiently about such problems as finding the complete subgroup lattice for a group and constructing homomorphisms or finding groups and subgroups with prescribed properties.

- Forming conjectures about finite group properties to eventually either prove or construct counterexamples.

Before considering using *FGB* for homework or lab assignments, an important note of caution is the following: Assignments must be chosen with thought and presented with proper guidance. Some of the more popular *ESG* or *FGB* labs ask students to perform tasks such as finding the subgroup lattice for a given group or constructing homomorphisms with a specified kernel or image. We have found that in the presence of limited knowledge, even with the software, such tasks can appear to be an enormous amount of busy work. When working on problems such as these, it is important to emphasize that applying careful reasoning and considering previously-learned theorems reduces the busy work. Instructors must take great care in knowing what theorems and techniques they wish to introduce to students when starting a lab, and they must monitor student progress as the labs are completed. Each lab should then be followed with a discussion about what was discovered and what can be proved for use in the future. In particular, we always make it a point to demonstrate efficient solutions to problems at carefully chosen points within the course. The beauty of the software is that it can be used simultaneously to demonstrate theorems and encourage the discovery and proof of other theorems.

3   An *FGB* Primer

Those familiar with programs in a *Windows*-based environment and with the basics of group theory should find *FGB* self-explanatory and easy to operate. However, since beginning students do not have this expertise in group theory, and perhaps are not familiar with the *Windows* operating system, they may need help from their instructor. In order to facilitate this process, we briefly outline the features of the program as well as indicate some functionality that might not be so apparent.

Thanks to the efficient work of *MAGMA*, *FGB* comes equipped with a library that contains all the finite groups up to order 16, all the nonabelian groups up to order 40, and also the alternating group A_5 of even permutations on five symbols. The files containing the groups are organized by folders, mostly according to the group size. The group files are text files in a special format that can be read by *FGB*. The names of the files are given as 4-digit numbers. The first two digits are the size of the group and the second two are based on an ordering of the groups within this particular size. Cyclic subgroups (for sizes 16 or less) can be found in the cyclic folder and always end in 01. For all groups of size 16 or less the numbering is identical to that used by Geissinger in *ESG*. From our web site [5], one can download a single file to install *FGB* either on a hard drive or a diskette. Installation on a diskette may be useful for the student who does not have a computer but does need a version of the library that they can alter or augment (especially in the Notes tab–see below).

Along with the classification theorem for finite abelian groups, the completeness of this library allows the user to ask and definitively answer questions about the existence of certain features within small groups. (The center button under the `Commutativity` tab can be used to quickly determine if a group is abelian or not.) For example, with a quick examination of 2001, 2002, and 2003 (the only groups of order 20 in the library) it is immediate that all nonabelian groups of order 20 have centers of order two or less. As any student of beginning group theory will quickly realize, studying groups of various orders often involves studying what kinds of subgroups and factor groups can be formed from a given group. As we will shortly explain, the software allows the user to extract subgroups and factor groups (when appropriate) and easily create new group files for later analysis. The completeness of the group library means that an instructor can always be confident in asking students to identify any group that arises in this way (i.e., specify the group in the library or give an abelian direct sum decomposition that represents the isomorphism type).

Each group to be considered is opened as a separate window within *FGB*. When many groups are opened simultaneously these can be arranged in a tiled format or on top of each other. Each group in the library can be viewed in a form where the elements are abstract symbols as in *ESG* (called standard names) or in a form where the elements are named according to their functionality (called descriptive elements). For smaller groups (of approximately size 16 or less) this descriptive format can involve a variety of possibilities including a type of direct sum notation, integers for modulo arithmetic, or permutations as in the case of S_3. The descriptive names for larger groups are given in terms of their generators and relations. (Always consult the `Notes` tab for these details with any particular group.)

Drag and edit features give the user the ability to reorder the elements in a group or rename them in a more suitable form. One can also enter one's own groups with the program. This approach is necessary if one needs an abelian group of order more than 16 or if there is an abstract group that needs to be identified within the library.

When a group table is opened, in addition to the usual menus of `File`, `Edit`, `Window`, and `Help` (which reside within the shell window of *FGB* and thus not with any particular group), the following tabs are available: `Axiom Check`, `Group Table`, `Subgroup Table`, `Commutativity`, `Homomorphisms`, and `Notes`. The function of each of these is briefly outlined below.

3.1 File, Edit, Window, and Help Menus

The `File`, `Edit`, `Window`, and `Help` menus provide the usual *Windows* functions with some features specific to group theory. The `File` menu allows one to open or create new group files. The usual self-explanatory capabilities of `Save`, `Save As` (for the group files), and `Exit` are also available. The `Edit` menu allows one to modify existing tables by adding, removing, or renaming elements. This tab works on the window currently considered active. The `Window` tab (or just clicking on the desired window) is a method of changing the window that is currently active. The `Window` menu also has a command to tile all the open windows in an attempt to see as much as possible of all of them simultaneously. This can be particularly useful for in-class demonstrations to show the differences in groups of small order (such as the groups of order four or eight). At the time of this writing, the `Help` menu is only used to remind the user to register the software and to provide information about the authors.

3.2 Axiom Check

The `Axiom Check` tab provides a means, by using the Cayley table, to verify that we are working with a group. Tabs for checking for an identity, inverses for all elements, and associativity are available. (The inverse check can also be run on a group just to produce a list of elements and their inverses for later reference.) When any of these operations fail, the user is given appropriate feedback. This provides the student with the opportunity to realize how special the structure of a group is and how surprising it is that there are so many groups (with impressive variety) of a modest size. Note: The original DOS program of *ESG* came with a library of non-group Cayley tables available for study. While this is not currently included with this version of *FGB*, it would not be hard to incorporate if there was suitable demand.

3.3 Group Table

The `Group Table` tab provides a way to examine the Cayley table of a group. In addition to the functions available under the `Edit` menu described above, it is from here that the group table can be reordered, relabeled, or altered on an entry-by-entry basis. Among other advantages, these features allow instructors to construct modified Cayley tables for use on exams and quizzes as well as to enable students to see how sensitive the axioms can be to changes as small as altering a single entry within the table.

3.4   Subgroup Table

The `Subgroup Table` tab can be used to choose which elements from a group are to be used in forming a subgroup. The program then automatically adds the necessary elements in order to form the closure, incorporating the user's choices. The size of the subgroups formed is always indicated in the upper left corner of the table for the subgroup. This means that when cyclic subgroups are formed, the order of elements can be easily identified. Subgroups can always be conjugated by any element from the original group. Once a subgroup is formed, the group can be exported to its own file window and later saved or modified for future use. One can also view the cosets of a subgroup. This feature then reorders the table so that elements in the same coset appear contiguously. The program employs a coloring scheme that works best when there are no more than 16 cosets, since each coset is assigned a unique color. As originally conceived by Geissinger in *ESG*, when the subgroup is normal, this coloring scheme indicates that the induced operation on cosets is well-defined. If the subgroup is normal, it then becomes possible to export the quotient group for further analysis in its own window. (The user should note that when exporting a coset table, the entries are replaced by a consistent representative of each coset, thus producing a Cayley table like the others in the library.) When subgroups are not normal, the conjugation option helps remind students that conjugates of nonnormal subgroups also yield nonnormal subgroups. Of course when they are normal, conjugation only acts to (possibly) reorder the elements.

3.5   Commutativity

The `Commutativity` tab can be used to calculate the center and centralizers of the elements in a group. When nontrivial, the center is always at least one interesting example of a normal subgroup and hence a uniformly colored coset table and resulting factor group. In *FGB* 2.0, if any of these groups are desired for export or factor group formation, one must manually recreate them with the subgroup tab. This is not usually difficult since ordinarily a small subset of the elements in the group will generate the entire subgroup.

3.6   Homomorphism

The `Homomorphism` tab is where homomorphisms can be constructed between any two open group windows. To do this, one operates in the window designated as the domain and then selects from this the codomain. The homomorphisms are specified in pieces in the following sense. After assigning the function's image for an element from the domain, the program indicates what this implies about the function on the rest of the elements in the subgroup. This process can proceed step-by-step or it can be done quickly once the student understands how things work. In many cases, the initial specifications will lead to a contradiction and this is indicated by the software. When this happens, it is possible, without starting over completely, to back up to a previous stage of the definition where no contradiction existed. A table format is used to display the homomorphism. This is similar to the original Cayley table for the group or a subgroup except that the header columns and rows are two squares thick to indicate both the domain and image element for the homomorphism. The body of the table then shows the products of image elements. This reminds students about the group-nature of the image as well as allowing them to see consequences, such as the fact that every element in the image has preimages of the same size.

3.7   Notes

The `Notes` tab contains background information about the group (such as the common name for the group, format for the descriptive elements or generators and relations) as well as providing a place to make notes about one's investigations of a group.

As we have previously mentioned, for groups over size 16, the presentation and other group information was provided by *MAGMA* using $x \wedge y$ to denote $y^{-1}xy$ and (x, y) to denote $x^{-1}y^{-1}xy$. A description of some groups also appears when the group has a common identification such as dihedral, symmetric, alternating, or it can be written as the direct sum of smaller (possibly involving abelian or even cyclic terms) groups. In these identifications, `Dihedral(n)` is used to denote the group of symmetries of the regular polygon with n sides. Occasionally, such products may also involve notation such as `GroupOfOrder(12,3)`. This is *MAGMA*'s own internal notation for some group of order 12. (A good exercise might be to determine, with justification, exactly which group of order 12 in the library this is and explain why.) Since newer versions of *MAGMA* no longer provide this identification information (now only the presentation is available), it is no longer possible to make effective systematic use of such notation.

When working with the `Group Table` tab, clicking with the right mouse button gives the option of copying the entire table to the notes page for printing later. These tables can then be pasted from the notes to any spreadsheet, where it is possible to format and print them as desired.

When looking for subgroups with the `Subgroup Table` tab, a right click provides the option of copying the list of elements in the subgroup to the notes. When this happens, the elements are listed in alphabetical order. This has the advantage of telling one at a glance whether subgroups produced from different generators within a given group are equal or not.

4   Example Problems

In order to give an idea as to how learning with *FGB* might take place, we offer some examples for using the software. This is not meant to be an exhaustive study of the possibilities but rather an illustration of how the concepts of beginning group theory can be addressed with this program. While we feel that more is now possible with *FGB* than with *ESG*, the genesis for many of these ideas is certainly due in part to Geissinger's original work and the materials of Parker.

In what follows, the solutions are idealized in the sense that they convey the maximum information that we would like the student to obtain from the problem. Without proper guidance and follow up, many students will glean far less and by more inefficient means.

4.1   Example 1. Finding Subgroups

Problem. Up to isomorphism, how many subgroups of order 8 does the group $2411 = \langle a, b, c, x \mid a^2 = b, b^2 = c, c^2 = 1, x^3 = 1, x^a = x^2 \rangle$ have? (Here, for x and y in a group, x^y is defined as $y^{-1}xy$.) What are these isomorphism types?

Solution 1. Examination of the library for groups of order 8 indicates that there are five different groups of order 8. Note that from any Cayley table for a group, one can count the number of elements of order 2 as the number of times, excluding the identity, where the identity appears on the diagonal. A brief study reveals the following groups of order 8.

0801	Cyclic $\mathbb{Z}_8$ —	one element of order 2 and two elements of order 4
0802	$\mathbb{Z}_4 \oplus \mathbb{Z}_2$ —	three elements of order 2
0803	$\mathbb{Z}_2 \oplus \mathbb{Z}_2 \oplus \mathbb{Z}_2$ —	seven elements of order 2
0804	D_4 (symmetries of the square) —	five elements of order 2
0805	The Quaternions —	one element of order 2 and six elements of order 4

By checking the diagonal of 2411 we see that c is the only element of order 2. Therefore, 2411 cannot have subgroups isomorphic to any of 0802, 0803, or 0804. When either using the `Subgroup Table` tab to check orders or by searching on the diagonal for occurrences of c, it is not hard to see that 2411 has only two elements of order 4. This means that no subgroup of type 0805 is possible and hence the only possible subgroups of order 8 from 2411 are cyclic. By using the `Subgroup Table` to examine the subgroups produced from single elements, it is readily apparent that such cyclic subgroups do indeed exist so the answer to the original question is that there is only one subgroup of order 8.

Solution 2. From the Sylow theorems, we know that since 2411 has size 24 it contains a subgroup of order 8 and all subgroups of order 8 are conjugate (and hence isomorphic) to one another. Thus we immediately know that the answer to the question is one and we can quickly see from the `Subgroup Table` tab that such subgroups are cyclic.

Comments. It is our experience that a first-semester course would probably not cover the Sylow theorems, but they are certainly a possibility for subsequent courses as well as a check for the instructor that a correct solution has been found. Despite this, it is easy to find examples of this type of problem (e.g., subgroups of size 8 in groups of order 16 or 32) where the only line of reasoning available is more like that in the first solution and the final answer could be different than one. This affords many versions of this type of problem since the library has 14 groups of order 16 and 44 groups of order 32. In the general setting, the solution to a problem like this can become quite complicated, forcing students to specifically attempt to construct the various subgroup types of order 8 within the larger group. Knowing that elements of order 4 come in pairs (i.e., x has order 4 iff x^{-1} has order 4) or that some of the groups of order 8 (i.e., 0802 and 0804) have varying numbers of noncyclic subgroups of order 4, are just some examples of the kinds of observations that can prove to be useful in deciding which combinations of elements need to be tried (with the `Subgroup Table` tab) to produce the various groups of order 8.

4.2   Example 2. Factor Groups

Problem. Prove or disprove the following conjecture. Suppose that H and K are normal subgroups of a finite group G. If H and K are isomorphic, then G/K and G/H are isomorphic.

Solution. The conjecture is false and can be illustrated by many counterexamples. In addition to realizing that they cannot construct a proof, the problem for students is to locate such an example. Part of the frustration here is that most subgroups of the groups they want to consider aren't normal and therefore won't yield quotient groups. One clever approach to circumvent this difficulty is to consider groups with large centers. If G is a nonabelian group with center H, then any subgroup K of H will be normal in G and will thus always yield a quotient group G/K. Consider, for example, the group $3201 = \langle a, b, c, x, y \mid a^2 = x, b^2 = y, b^a = bc \rangle$. The center of 3201 is a subgroup of order 8 generated by c, x, and y, each having order 2. Thus $\{1, x\}$ and $\{1, c\}$ are isomorphic, but $3201/\{1, x\}$ and $3201/\{1, c\}$ are not. We know this since these two groups of order 16 do not have the same number of elements of order 2: $3201/\{1, x\}$ has seven elements of order 2 while $3201/\{1, c\}$ has only three.

4.3   Example 3. Homomorphisms

Comment. Problems such as the following force the student to grapple with the fundamental isomorphism theorem that states for $\varphi : G \to H$ then $G/\ker(\varphi) \cong \varphi(G)$. We encourage instructors to illustrate this theorem in class as follows. The program allows one to construct homomorphisms between any two groups. One can then use the `Subgroup Table` tab to create the factor group of the domain modulo the kernel. (One receives confirmation here that the kernel is indeed normal.) This factor group can be displayed along with the image group (which will appear with the `Homomorphism` tab). For reasonably sized examples, when these two windows are shown side-by-side, the isomorphism is apparent. However, one can explicitly construct the isomorphism by first properly exporting the factor group and, if φ is not onto, the image subgroup in H. The `Homomorphism` tab on the exported factor group window is then the place to formulate the isomorphism, that, as should be pointed out, is canonical to define.

Problem: Construct a homomorphism from $G = 2404 = \langle a, b, c, x \mid a^2 = b^2 = x, x^2 = c^3 = 1, b^a = bx \rangle$ onto a group of order 4. Does it matter which group of order 4 you use? Explain.

Solution: There are exactly two groups of order 6 and two groups of order 4:

0601	$\mathbb{Z}_6$	0401	$\mathbb{Z}_4$
0602	D_3	0402	$\mathbb{Z}_2 \oplus \mathbb{Z}_2$

Suppose that $\varphi : G \to H$. Since $|H| = 4$, and φ is onto, the fundamental homomorphism theorem tells us that we must have that $|\ker(\varphi)| = 6$.

Since 0602 has three elements of order 2 but G has only one element of order 2 (i.e., x), G cannot contain any subgroups isomorphic to 0602. Checking the orders of the elements in G reveals that there is exactly one cyclic subgroup of order 6 (i.e., that generated by cx). Since this is the only such subgroup, it is fixed by inner automorphisms and hence must be normal. Therefore, the image of φ must be isomorphic to $G/\langle cx \rangle$. Looking at the coset table for this factor group we see that it has three elements of order two and thus must be isomorphic to 0402. This means that there is no such homomorphism onto 0401, so it does indeed matter which group of order 4 is chosen.

The software then shows that the homomorphism can be specified by letting $\varphi(cx) = 00$, $\varphi(b) = 10$, and $\varphi(ax) = 01$. (The software uses 00 for the pair $(0,0)$ and so on.) The secret to finding a correct specification is that we let the kernel be $\langle cx \rangle$ and then we note that elements from distinct cosets of $\langle cx \rangle$ must go to distinct (nonidentity) elements in 0402. In fact, any choice where b and ax go to distinct elements of order 2 in 0402 will work.

4.4 Example 4. The Sylow Theorems

Comment. The software can be used in many ways to teach the Sylow theorems. As in Example 1, a knowledge of these theorems can be useful in solving many subgroup or isomorphism-type questions for the groups in the library. Alternatively, after learning the statements of the theorems (and perhaps their proofs), the software can be used to illustrate their conclusions. Here we give yet a third possibility of presenting a series of problems whose solutions force the student to examine concrete examples of the concepts involved in the Sylow theorems and to construct simplified proofs for special cases of those theorems (see, for example, the presentations given in [2]). This should ease the way toward a more general proof. We believe that this method of teaching a proof via simplified examples has other useful realizations with this program.

Setting. Consider the group $2001 = G = D_{10} = \langle a, b, c \mid a^2 = b^2 = c^5 = 1, c^a = c^4 \rangle$. Recall that this is the group of symmetries of a regular decagon. Use G to answer the following questions.

Problem 1. Partition G into conjugacy classes and write the class equation for G.

Comments. Recall that for $a \in G$, the centralizer of a, $C(a)$ (which can be calculated from the Commutativity tab), is the subgroup of all elements in G that commute with a. It is a standard fact that the number of elements in the conjugacy class of a is the index $[G : C(a)]$ of $C(a)$ in G. The class equation is then just a way of accounting for every conjugacy class in G by the formulation $|G| = \sum_i [G : C(a_i)]$ where the sum involves exactly one element from each conjugacy class. These results and notions could easily be presented in class or homework before questions like these are investigated.

This problem is also an example where the power of the software has been suitably limited to force the beginning student to think more and push buttons less. Since the program only has methods for conjugating subgroups and not conjugating single elements, the student must use ingenuity to do the calculations efficiently.

Solution. Since conjugate elements have the same order, we first partition G into subsets corresponding to the orders of the elements. This produces the following partition: $\{1\}$ (order 1), $\{a, b, ab, ac, ac^2, ac^3, ac^4, abc, abc^2, abc^3, abc^4\}$ (order 2), $\{c, c^2, c^3, c^4\}$ (order 5), and $\{bc, bc^2, bc^3, bc^4\}$ (order 10).

From the facts cited above about $C(x)$, we know that the size of each conjugacy class must divide $|G| = 20$. Of course $\{1\}$ is a conjugacy class and the fact that the center of G consists of $\{1, b\}$ immediately tells us that $\{b\}$ is the only other singleton conjugacy class. For the elements of order 10, we note that these must divide into two classes of size 2 since, for example, $|C(bc)| = 10$. Choosing an element, say a, not belonging to $C(bc)$ and computing $a * bc * a^{-1} = a * bc * a = abc * a = bc^4$ shows us that $\{bc, bc^4\}$ and $\{bc^2, bc^3\}$ are these two conjugacy classes. Similarly, $|C(c)| = 10$ and so with $a \notin C(c)$ we have that $a * c * a^{-1} = c^4$ so $\{c, c^4\}$ and $\{c^2, c^3\}$ give the two conjugacy classes for elements of order 5.

To compute the conjugacy classes of the elements of order 2 we can reason as follows. Since $|C(a)| = 4$, the conjugacy class of a has size 5. Furthermore, conjugating a in turn by abc, abc^3, abc^4, bc^3 (one element from each coset of $C(a)$) lists for us that the conjugacy class of a is $\{a, ac^2, ac, ac^3, ac^4\}$. Having already accounted for b above, dimension considerations immediately tell us that $\{ab, abc, abc^2, abc^3, abc^4\}$ is the remaining conjugacy class. Therefore the class equation becomes $20 = 1 + 1 + 5 + 5 + 2 + 2 + 2 + 2$. When computing various conjugates of a it is helpful to be able to return quickly to the subgroup $\{1, a\}$ after each conjugation. One can do this by conjugating by the inverse of any element used in the process. The Axiom Check tab can be used to provide a list of elements and their inverses for easy reference.

Problem 2. Suppose G is any group (abelian or nonabelian) with the above class equation. Must such a group have a subgroup of order 5? Explain.

Solution. Yes. Each term on the right-hand side of the class equation is the index of a centralizer in G so if we divide $|G| = 20$ by these numbers then we get the order of the centralizers, which are subgroups of G. Thus, since there are conjugacy classes of size 2 there must be centralizers of size 10. A check of the groups of size 10 (there are only two: $1001 = \mathbb{Z}_{10}$ and $1002 = D_5$) shows that every group of size 10 always contains a subgroup of size 5.

Problem 3. In the group 2001, how many subgroups of order 4 are there and what isomorphism types occur?

Solution. The check of orders made in part (a) shows that G has no elements of order 4 and thus the only possible subgroups of order 4 are of type 0402 ($\mathbb{Z}_2 \oplus \mathbb{Z}_2$). Such a subgroup will be formed from three elements of order 2. Since 2001 has eleven elements of order two, we do not want to attempt to find all such subgroups simply by trial and error. However, it is reasonable to use trial and error to find one of these, say $H = \{1, a, b, ab\}$. Furthermore, we can now look for other such subgroups by conjugating this one. When we do this, we notice that all conjugations by elements not belonging to H give something besides H (and thus, the normalizer $N(H) = H$). Furthermore, we can easily realize that the same is true for conjugates of H; i.e., $N(gHg^{-1}) = gHg^{-1}$ for any $g \in G$. Thus there are $[G : N(H)] = \frac{20}{4} = 5$ order 4 subgroups conjugate to H.

We claim that this collection of subgroups $\mathcal{H} = \{H = H_1, H_2, H_3, H_4, H_5\}$ is the entire list of order 4 subgroups of G. Suppose that $W = \{a_0 = 1, a_1, a_2, a_3\}$ is any order 4 subgroup of G. Since this is a copy of 0402, we know that if $1 \le i, j, k \le 3$ are all distinct then $a_i a_j = a_k$. Given a specific H_i, we can consider all the conjugate subgroups $a_j H_i a_j^{-1}$ for $j = 0, 1, 2, 3$. These will all belong to $\mathcal{H}$ and the process of doing this for all i and j will partition $\mathcal{H}$ into a collection of disjoint subsets where two subgroups in $\mathcal{H}$ belong to the same subset iff they are conjugate in this way by an element from W.

We claim that each such subset (called a W-orbit) will have either one, two, or four elements. Let $\langle H_i \rangle$ denote the W orbit of H_i from $\mathcal{H}$. From what we said above, we know that $a_j H_i a_j^{-1} = H_i$ iff $a_j \in H_i$. Thus, $|\langle H_i \rangle| = 1$ iff $H_i = W$ and $H_i \cap W = \{1\}$ iff $|\langle H_i \rangle| = 4$. Since the order of $W \cap H_i$ must divide 4, we know that this intersection cannot have size 3. Thus the only remaining possibility is when (without loss of generality) $H_i \cap W = \{1, a_1\}$. In this case we have that $\langle H_i \rangle = \{H_i, a_2 H_i a_2^{-1}, a_3 H_i a_3^{-1}\}$. That this set has size 2 follows once we note that $a_2 H_i a_2^{-1} = a_3 H_i a_3^{-1}$. This is clear because conjugation (which is injective on subgroups) of both $a_2 H_i a_2^{-1}$ and $a_3 H_i a_3^{-1}$ by a_2 gives H_i (observing that $a_2^2 = 1$ and $a_2 a_3 = a_1$).

Therefore $\mathcal{H}$, which has size 5, is partitioned into disjoint W-orbits of sizes 1, 2, and 4. Therefore, there must be an orbit of size 1. It is the H_i in this orbit that must equal W. Thus, there are exactly five subgroups of order 4, they are all of type 0402, and they are all conjugate to one another.

5   Plans for the future

Based on the registration information we have received, it now appears that approximately 40 schools throughout the USA and Canada have expressed an interest in using *FGB*. In addition to this, if future *Windows* environments stop supporting the DOS 16 bit format used by *ESG*, then *FGB* will become an increasingly important alternative for those who wish to teach in this manner. It is our sincere hope that *FGB* can live up to the expectations of those educators who will want to use it in the future. We are trying to learn what new features need to be added as well as what existing bugs need to be fixed. Based on the comments we have received so far, we have constructed the following list of most commonly considered improvements. These are listed from the most routine to the most ambitious.

- The ability to see the formation of subgroups in a step-by-step fashion but without sacrificing the speed that currently exists for those who don't need this.

- The option to record homomorphism formulas in the notes by right clicking as in the `Group Table` and `Subgroup Table` tabs.

- More on-line documentation for using the program, especially intended for the beginning student.

- The addition of a calculator tab to each group window. This would allow the user to make calculations in the group and solve equations.

- The ability to calculate the orders of elements more directly.

- Inclusion in the library of all abelian groups of size 40 or less and possibly Cayley tables for sets that are not groups.

- A web-based version of *FGB*. This would eliminate the need to use a *Windows* platform for *FGB* and thus make the software available to anyone with access to the internet.

- The ability to perform constructions with groups such as direct sums or even semi-direct products.

- A graphical interface that would allow the user to grasp the geometric interpretations of some of the groups (such as the dihedral groups), when this is appropriate.

- Improved connections of *FGB* to the applied areas of group theory, especially cryptography and combinatorics.

- Automated capabilities to search the library for common features of groups or the occurrence of specified phenomena.

For students, one of the excellent features that programs such as *ESG* or *FGB* offer is the opportunity to survey the features of a wide variety of groups in an attempt to find their similarities and differences. These similarities can lead to conjectures and subsequent research by the student. These aspects are a big part of the teaching style exemplified in [8] or the results obtained in REU projects such as those of [1, 7]. We have been told it would be helpful to automate this process and look for various features of many groups in the library and keep statistics on what is found. Thus, some ability to write and perform macros on one or more groups in the library might be useful. However, it is not at all clear if enough functionality would be possible and since *MAGMA* and *GAP* currently have this capacity, incorporating it may not be worth the effort.

The authors are proud of *FGB* and we feel strongly that both in its current form and in future releases, it has a great potential to enhance the beginning group theory classroom. We sincerely hope that, after a careful examination of this paper and use of the program, the reader will agree. However, for *FGB* to continue to grow effectively, we depend heavily on user input. Please write to Ed Keppelmann at *keppelma@unr.edu* with any comments, questions, wishes, and concerns. Given the existence of continued demand for the software, new releases of *FGB* may occur more frequently. We are committed to making sure that the software always remains free and relevant for those individuals who wish to use it.

References

[1] S. Belcastro and G. Sherman, *Counting centralizers in finite groups*, Mathematics Magazine **67** (1994), no. 5, 366–374.

[2] T. Hungerford, *Abstract Algebra, An Introduction*, 2nd ed., Saunders College Publishing, 1997.

[3] The GAP Group, *GAP — Groups, Algorithms, and Programming*, Version 4.2; Aachen, St Andrews (`http://www-gap.dcs.st-and.ac.uk/gap`).

[4] L. Geissinger, *Exploring Small Groups: A Tool for Learning Abstract Algebra*, now only available bundled with [8].

[5] E. Keppelmann and B. Webb, web site for *Finite Group Behavior*, `http://unr.edu/homepage/keppelma/fgb.html`.

[6] *MAGMA*, Computational Algebra Group of the School of Mathematics and Statistics, University of Sydney, `http://www.maths.usyd.edu.au:8000/u/magma/`.

[7] C. Mitchell and G. Sherman, *Counting products of triples in finite groups*, Proceedings of the 26th Southeastern Conference on Combinatorics, Graph Theory and Computing, *Congressus Numerantium* **109** (1995), 129–134.

[8] E. Parker, *Laboratory Experiences in Group Theory: A Manual to be used with Exploring Small Groups*, Mathematical Association of America, Washington, DC, 1998.

Edward Keppelmann, Department of Mathematics AB601 MS084, University of Nevada Reno, Reno, Nevada 89557;
keppelma@unr.edu.
Bayard Webb, Department of Mathematics AB601 MS084, University of Nevada Reno, Reno, Nevada 89557;
keppelma@unr.edu.

Discovering Abstract Algebra with *ISETL*

Ruth I. Berger

Abstract. *ISETL* stands for Interactive SET Language. It is a mathematical programming language whose syntax resembles standard mathematical notation. In this paper I describe how I use *ISETL* in my abstract algebra class at Luther College. Many examples of *ISETL* programming code are included, so interested instructors can easily evaluate the usefulness of this tool for their classes.

1 Classroom Experience

At Luther College we offer abstract algebra each fall semester with an enrollment of 15 to 20 students. Since the fall of 1995, my abstract algebra classes have met in a computer classroom. Each week the students spend one or two class periods working in groups on worksheets that I have developed. These worksheets use *ISETL* to let the students experiment with many examples. They come up with conjectures before we prove the corresponding theorems in class. The students develop a very good feeling for algebraic concepts and they feel comfortable with a wide range of examples. A good illustration of this is the day when we collect all the groups of order 4 that we have seen. I am always very impressed with the wide variety of replies students come up with; see Table 1 for some examples. (Here we use $+_m$ and $\cdot_n$ to denote addition mod m and multiplication mod n, respectively.) We then have a class discussion about which groups "look alike," up to renaming and reordering of the elements. This leads us to discovering the concept of isomorphic groups.

$$(\mathbb{Z}_4, +_4) \qquad (\{5, 25, 35, 55\}, \cdot_{60}) \qquad (\{1, -1, i, -i\}, \cdot)$$
$$(U(10), \cdot_{10}) \qquad (U(8), \cdot_8) \qquad (\{(1), (12), (34), (12)(34)\}, \circ)$$
$$(\mathbb{Z}_5 - \{0\}, \cdot_5) \qquad (\{1, 3, 7, 9\}, \cdot_{20}) \qquad (\{0, 3, 6, 9\}, +_{20})$$

Table 1: Some Groups of Order Four

I like this approach to teaching abstract algebra. My students develop an ownership of the mathematics involved, which is not commonly seen in an undergraduate mathematics course. One might think that it is somewhat time consuming to have the students learn *ISETL*, but they are in fact learning correct mathematical notation and reasoning. The challenge of having to "explain" to the computer exactly what is intended is a great learning experience. Unlike the instructor, the computer gives no partial credit and no amount of hand waving or "You know what I mean" will get the desired result. The students have to thoroughly understand the underlying mathematics in order to construct *ISETL* programming code that yields the desired result.

The textbook we use is Gallian's *Contemporary Abstract Algebra* [3]. My worksheets serve as an introduction to new material; the students then read the corresponding sections in the book. Some proofs are discussed in class and both individual and group homework is assigned from the text book.

Our computer classroom is set up for two students to share a computer. This encourages student interaction in our classes. The tables are arranged in a horseshoe shape with the monitors close to the wall, facing the inside of the room. An instructor standing in the middle of the room can easily survey the progress of all students. Students sitting at the computers can swing their chairs around to another set of tables, also arranged in a horseshoe shape, which enables them to face each other and the white board. The class mode can alternate between class discussion, computer exploration, and work on the white board.

The students work in groups of three or four, though I prefer groups of three. Sometimes a group of four works as a loose association of two pairs who get together only for a final write-up or when the pairs are completely stuck. Occasionally a group of four is crowded around one monitor, and I split that group up into two pairs. Groups of three usually function well. Sometimes I need to remind them to alternate the person on the keyboard. When I first started teaching the course this way, a few students had no programming experience at all. They had a difficult time writing the relatively simple programming code using *ISETL*. In recent years, students have become much more comfortable with using computers and this is no longer a major obstacle.

I should mention that while most of my worksheets use *ISETL*, we also use *Exploring Small Groups* [2]. It has a nice feature that allows one to rename and reorder the elements in a Cayley table. Its best feature, however, is its color-coded presentation of operations on sets of cosets.

I am very thankful to Ed Dubinsky and Uri Leron for introducing me to *ISETL* at a workshop at Purdue University in the summer of 1994. Several of the *ISETL* program codes in this paper were inspired by their book [1]. Another interesting paper to read in this context is their paper [4]. Details for downloading *ISETL* can be found in the appendix and at the web site for this volume.

2   Sets and Functions

The most basic *ISETL* structures are sets and functions. *ISETL* can only handle finite sets. Set builder notation is used in the following way: {*elements : domain | property*}. Similarly, ordered tuples can be constructed, where repetition of elements is allowed. Note, however, that the command `!setrandom off` will cause the elements of a set of numbers to be displayed in an increasing order. Here are some examples of *ISETL* code.

```
S := {6,8,2,17,13,5};   Z40 := {0..39};   T := [5,4,4,1,7];
S3 := {[a,b,c] : a,b,c in {1..3} | #{a,b,c}=3};
C := {[x,y,(x*y) mod 6] : x,y in {0..5}};
m40 := func(x,y); return (x*y) mod 40; end;
U40 := {g : g in Z40 | (exists g' in Z40 | g' .m40 g =1)};
```

If a set G is defined, the commands `forall x in G` |*property* and `exists x in G` |*property* cause *ISETL* to reply with `true` or `false`. The command `choose x in G` |*property* causes *ISETL* to randomly choose an element of G having the required property. If there is not an element with the required property, *ISETL* returns om.

ISETL has several different function notations. The function that squares each entry can be written either as `f := func(x); return x**2; end;` or as `f := |x -> x**2|;` (`**` is used for exponentiation). The first notation allows one to specify a domain:

```
f := func(x); if x in {0..10} then return x**2; end; end;
```

Furthermore, *ISETL* recognizes a set of ordered pairs as a function. For an example, see the section on isomorphisms below.

An excellent challenge for students is to get them to reflect on the concepts involved by writing *ISETL* code that checks whether a function is one-to-one or onto. Possible solutions are:

```
one1 := func(f,S1); return forall a,b in S1 | a/=b impl f(a) /= f(b);
   end;}
one1 := func(f,S1); return forall a,b in S1 | f(a)=f(b) impl a=b; end;}
onto := func(f,S1,S2); return forall y in S2 | (exists x in S1 |
   f(x)=y); end;}
```

As hinted above in the first examples of code, the function implementing multiplication modulo 40 can be defined as follows.

```
m40 := func(x,y); return (x*y) mod 40; end;
```

Evaluating this function requires two inputs, such as in `m40(5,4);`. Another way of evaluating this function (using infix notation instead of prefix) is `5 .m40 4;`. After students have been introduced to this notation, I have them work on the following discovery exercises in class. (In what follows, "discovery" exercises denote in-class work where an instructor is available for feedback, supervision, and direction. In contrast, an "assignment" is homework to be done outside of class.)

Discovery. Is m40 a binary operation on Z40? Is m40 a binary operation on Z40-{0}?

Discovery. Use *ISETL* to examine properties of the set $\{0..n-1\}$ under multiplication modulo n. In particular, determine which elements have inverses. Start with $n = 13$, define G := {0..12}; and o := |x,y -> (x*y) mod 13|;. Repeat your investigation for other values of n. What do you notice? Note: You just have to define the new G and o. Then you can copy the same commands.

Some examples of acceptable programming for the second discovery investigation include the following.

```
exists x in G | (forall g in G | x .o g = g);
exists x in G | (forall g in G | x .o g = g and g .o x = g);
e := choose x in G | (forall g in G | x .o g = g);
forall g in G | (exists g' in G | g' .o g =e and g .o g' = e);
{g : g in G | (exists g' in G | g' .o g =e)};
{{g,g'} : g,g' in G | g .o g'=e};
```

These are more sophisticated commands and are usually obtained by my prompting an individual group on how they can improve their current code. If a group has particularly nice code, I sometimes ask them to share it with the rest of the class. However, at times other groups may not have reached the level of understanding where they are ready to appreciate it. It is better to let the "slower" groups struggle at their own pace rather than let them have code whose construction they do not understand.

Once students have made a conjecture on which elements of $\mathbb{Z}_n$ have multiplicative inverses modulo n, it is of course necessary for them to give a proof. While it is easy to see why elements that are not relatively prime to n can not have multiplicative inverses, it is much harder to verify the existence of inverses for the other elements. For this, one can use the Euclidean Algorithm. I do not expect my students to be able to come up with the code that implements this algorithm. Instead, I give them the following recursively defined function and ask them to investigate its purpose. This can be a little challenging for average students. It needs to be combined with having them do the algorithm "by hand" for a few smaller examples.

```
Euc_alg := func(a,b);    if (a mod b) = 0 then return b;
   else writeln a, ''='', a div b, ''*'', b, ''+'', a mod b ;   end;
   return Euc_alg(b, a mod b);   end;
```

Permutation groups and their subsets provide another important class of examples in an abstract algebra course. Here is how to define S_3 and S_4, using *ISETL*. The approach for S_3 looks nicer but takes more computational time than the method for S_4.

```
S3 := {[a,b,c] : a,b,c in {1..3} | #{a,b,c}=3};
S4 := {[a,b,c,d] : a in {1..4}, b in {1..4}-{a}, c in {1..4}-{a,b},
   d in {1..4}-{a,b,c}};
```

A composition on either of these sets can be defined in the following way. Note that here the function on the right is applied first.

```
os := func(p,q); if #p=#q then return [p(q(i)):i in [1..#p]]; end; end;
```

We also need A_4 and D_4 for future ISETL investigations. The following homework assignment shows students how to achieve these.

Assignment.

1. Save A3 := $\{[1,2,3],[3,1,2],[2,3,1]\}$; and A_4 (the set of even permutations of S_4) to a file to make them available for future computations.

2. Complete example #3 on p. 92 of our text, showing that D_4, the symmetry group of a square, can be viewed as a subgroup of S_4.

3. Define R1 := [2,3,4,1]; and H := [2,1,4,3];. Use these to compute the corresponding elements in S_4 for all elements of D_4. For example, $R2 = R1 \circ R1 = [2,3,4,1]$ .os $[2,3,4,1] = [3,4,1,2]$. Now make D_4 available for future *ISETL* computations by placing each of the above elements and D4 := {R0, R1, R2, R3, H, V, D, D'} into a file.

3 Group Properties

Once the students are familiar with a wide range of examples we can explore group properties. A common misconception at this stage of the course is that students expect the identity element to always be some version of 0 or 1. It is an interesting exercise to have them consider the set $\{5, 15, 25, 35, 45, 55\}$ under multiplication modulo 60. Here 25 is the identity. Closure and associativity hold, but not every element has an inverse. After removing the elements that are not invertible, all the group properties are satisfied and one obtains the Klein-4 group. Students investigate similar examples in the next discovery exercise. Note that this can be quite time consuming. After letting the students work on it for a while, the instructor probably needs to point out a few time saving alternatives (perhaps including the *ISETL* functions given later).

> **Discovery.** Table 2 lists sets and a corresponding operation. In each case, determine if each of the defining group properties are satisfied. If an identity element exists, state it explicitly. If any of the other properties are not satisfied, indicate a counter-example. Many of these can be best examined by just "looking" at them, but others may require some computation (perhaps with *ISETL*). Use your judgment as to which approach is best.

$$(\{1, 3, 7, 9, 11, 13, 17, 19\}, \cdot_{20}) \quad (\mathbb{Z}_{20}, +_{20}) \quad (\mathbb{Z}_{20} - \{0\}, \cdot_{20})$$
$$(\{1, 3, 7, 9, 11, 13, 17, 19\}, +_{20}) \quad (\{1, 3, 7, 9\}, \cdot_{20}) \quad (\{1, -1\}, \cdot)$$
$$(\{4, 8, 16, 20, 28, 32\}, \cdot_{36}) \quad (\{1, -1\}, +) \quad (S_4, \circ)$$
$$(2\mathbb{Z}, +) \quad (\mathbb{Q}, \cdot) \quad (D_4, \circ)$$

Table 2: Which are groups?

After having written *ISETL* code that checks for group properties in various explicit examples, most students realize it is advantageous to define a function that can be applied to any set G and operation $\circ$. However, some students may still want to write new code each time a new set and operation is examined. I let the students proceed at their own pace, pointing out nice solutions that other groups are discovering. The groups each keep a disk with their own codes and functions. Below are illustrations of the general *ISETL* functions that are sought.

```
is_closed := func(G,o); return forall x,y in G | (x .o y) in G ; end;
is_associative := func(G,o); return forall a,b,c in G | (a .o b) .o
   c = a .o (b .o c); end;
has_identity := func(G,o); return exists e in G |
   (forall a in G | e .o a = a); end;
identity := func(G,o); return choose e in G |
   (forall a in G| e .o a = a); end;
has_inverses := func(G,o); e := identity(G,o); return is_defined(e)
   and forall a in G | (exists a' in G| a'.o a = e);  end;
inverse := func(G,o,a); e := identity(G,o); return choose a' in G |
   a' .o a = e; end;
is_group := func(G,o); return is_closed(G,o) and  has_identity(G,o)
   and has_inverses(G,o) and is_associative(G,o); end;
```

Note that the functions has_identity and has_inverses actually only check for a left-identity and left-inverses, respectively. This saves computing time, which can have an impact when groups of large order are examined. To be more accurate, one could modify these functions: in has_identity replace e .o a = a by e .o a = a and a .o e = a; in has_inverses replace a' .o a = e by a' .o a = e and a .o a' = e.

Other investigations that can be pursued follow.

> **Assignment.**
>
> 1. What does the following function do?
>
> ```
> order := func(x, G, o);
>    e := choose g in G | (forall a in G | g .o a = a);
>    b := x;   i := 1; while b /= e do b := b .o x; i := i + 1;
>    end; return i; end;
> ```

2. Given a list of examples of groups (such as in the previous discovery exercise), investigate the orders of the group elements. [Solution: use `for x in G do print x, order(x,G,o); end;`]

3. A group G is given together with a subset H. Determine if H is a subgroup of G. If it is not, explain why not. Use any method that is appropriate. [Solution: use `is_sub := func((H,G,o); if H subset G then return is_group(H,o); end; end;`]

4. Examine if a given group is cyclic. [Two solutions: use `(choose x in G | order(x) = #G)` or `(is_cyclic := func(G,o); return exists x in G | order(x)=#G; end;).`]

4 Isomorphisms, Cosets, and Lagrange's Theorem

In order to use *ISETL* to check whether a map has the properties required of an isomorphism, we must first enter in the map. To make this easier, we need to be aware that *ISETL* recognizes a collection of ordered pairs as a function. Rather than telling students this fact, I first have them work on the following short assignment.

Assignment. Let `f := {[1,6], [2,-2], [3,1]};` and predict what *ISETL* will return for each of the following: `f(1); f(2); f(3); f(6);`

We already have *ISETL* code that determines whether a function is one-to-one or onto. All that is needed now is code that checks whether a given map is a homomorphism. Let the operations on G_1 and G_2 be defined as o_1 and o_2, respectively. Students should by now be sophisticated enough to produce an *ISETL* function such as the following.

```
is_hom := func(f,G1,o1,G2,o2);  return forall a,b in G1 |
   f(a .o1 b) = f(a) .o2 f(b);  end;
```

They can put their code to use in the following assignment, but first they need to determine the map.

Assignment.

1. For each of the following pairs of isomorphic groups, determine an explicit isomorphism (function). Use *ISETL* to verify that your map has the required properties.

$$\{5, 25, 35, 55\} \cong \{R0, R2, H, V\} \qquad U(18) \cong \mathbb{Z}_6 \qquad \mathbb{Z}_7 - \{0\} \cong \mathbb{Z}_6$$

2. For each of the following groups, find a group to which it is isomorphic and give the isomorphism explicitly.

$$\mathbb{Z}_{11} - \{0\} \qquad U(22) \qquad \{x^2 \mid x \in U(29)\} \qquad (\{4, 8, 16, 20, 28, 32\}, \cdot_{36})$$

After introducing direct products, it is a good exercise to check students' understanding by having them write *ISETL* code that implements it. Here is how the Cartesian product can be obtained.

```
X := func(G1,G2); return {[a,b] | a in G1, b in G2}; end;
```

To define the operation on this set we need a function that inputs ordered pairs and then combines their first and second entries, respectively. Here is how this can be done, where `o1` is the operation of the first group and `o2` corresponds to the second.

```
ox := func(s,t); return [s(1) .o1 t(1), s(2) .o2 t(2)]; end;
```

With these definitions, students can then explore the fundamental theorem of finite abelian groups.

I introduce cosets with the worksheet shown below. It is quite a long discovery exercise, but it can be completed as homework. The students conjecture Lagrange's Theorem and they acquire all the necessary knowledge to put together a formal proof of it.

Worksheet for Cosets and Lagrange's Theorem

Definition. Let $(G, \circ)$ be a group with subgroup H. For $g \in G$, the (left) coset of g is the set $gH = \{g \circ h \mid h \in H\}$.

1. Write an *ISETL* function named `coset` such that for each element of G it returns its left coset with respect to a subgroup H.

2. For each group G and subgroup H given below, investigate the following. How many elements are in each of the cosets? How many different cosets are there? Which elements of G are in two or more cosets? Which elements of G are not in any coset? Are any of the cosets subgroups of G?

   (a) $G = \mathbb{Z}_{12}$ and H is the subgroup generated by 3.
   (b) $G = S_3$ and $H = \{(1), (12)\}$.
   (c) $G = D_4$ and H is the subgroup of rotations.
   (d) $G = U(32)$ and H is the subgroup generated by 9.
   (e) $G = \mathbb{Z}_{11}$ and H is the subgroup generated by 10.
   (f) $G = A_4$ and H is the subgroup generated by (123).
   (g) $G = A_4$ and $H = \{(1), (12)(34), (13)(24), (14)(23)\}$.

3. For each Group G and subgroup H from above, find the set of all left cosets, denoted G/H. How many elements are in G/H? Is there a relationship between $\#(G/H)$, $\#G$, and $\#H$?

The following is an example of a student-*ISETL* interaction that could occur in response to this worksheet. The student's input is preceded with > and the *ISETL* responses consists of lines not starting with >.

```
> coset := func(x); if x in G then return {x .o h | h in H}; end; end;
> $ #2 part a) [note that $ is a comment symbol]
> G :=  {0..11};
> H := {0, 3, 6, 9};
> o := func(x,y); return (x+y) mod 12; end;
> coset(2);
{2, 5, 8, 11};
```

After a few more commands of this type, students start to suspect that the same sets seem to appear all the time. Now is a good time to suggest that they start comparing intersections of these sets or to examine which ones are identical.

```
> coset(2) inter coset(4);
{ };
> coset(2) inter coset(5);
{2, 5, 8, 11};
> coset(2)=coset(5);
true;
> {b: b in G | 2  in coset(b)};
{2, 5, 8, 11};
```

Finally, if the group of students is comfortable enough with writing code, I might make a suggestion that helps the students write the following.

```
> forall a,b in G | coset(a) = coset(b) or coset(a) inter coset(b) = { };
true;
```

Other possible student-*ISETL* interactions on this worksheet include the following.

```
> #coset(2);
4;
> forall a in G | #coset(a) = 4;
true;
> GmodH := {coset(a) : a in G};
> GmodH;
{{0, 3, 6, 9}, {1, 4, 7, 10}, {2, 5, 8, 11}};
```

```
> #GmodH;
3;
> #G; #H;
12;
4;
```

After some class discussion to make sure that everyone has obtained the same conclusions, students are ready to come up with (or at least follow) a formal proof of Lagrange's theorem during the next class. An *ISETL* activity for the next week is given below.

Discovery.

1. Do your answers change if right cosets are considered instead of left cosets? For which of the examples is the partition of G into left cosets the same as the partition of G into right cosets?

2. Given a group $(G, \circ)$ and a subgroup H, define GmodH := {coset(g) | g in G}; and define an operation oo on GmodH, so that coset(a) .oo coset(b) can be performed. Use your is_group function to determine if G/H is a group under oo for the examples given.

3. In those cases where G/H is a group, find a group that is isomorphic to it.

For more details on how to expand on this, I refer to [1] or the next article by Karin Pringle in this volume. I prefer to use *Exploring Small Groups* [2] to let students explore visually how the set of cosets can sometimes be given a group structure.

As final reference for instructors, I include a problem set that is designed to let students discover the first homomorphism theorem, $G/\ker(f) \cong \operatorname{im}(f)$. The maps in part 1 were taken from [1, p. 129].

Discovery.

1. Which of the following maps are homomorphisms? Remember that for finite groups you can use your is_hom function.

   (a) $f : \mathbb{Z}_{20} \to \mathbb{Z}_5$ defined by $f(x) = x \pmod 5$
   (b) $f : \mathbb{Z}_{20} \to \mathbb{Z}_7$ defined by $f(x) = x \pmod 7$
   (c) $f : S_3 \to S_3$ defined by $f(\sigma) = (123) \circ \sigma \circ (132)$
   (d) $f : S_3 \to S_3$ defined by $f(\sigma) = (123) \circ \sigma \circ (123)$
   (e) $f : \mathbb{Z} \to \mathbb{Z}$ defined by $f(x) = 7x$
   (f) $f : \mathbb{Z} \to \mathbb{Z}$ defined by $f(x) = 7x + 2$

2. Write an *ISETL* function that returns the kernel of a homomorphism. Is the kernel always a group?

3. In determining the kernel of f, you computed the set of all elements $x \in G_1$ that are mapped to the identity element of G_2. Now, for each element $y \in G_2$, compute the set of all elements $x \in G_1$ such that $f(x) = y$. How many different sets do you get? For each set, how many elements does it have? Do you notice anything familiar happening?

5   Conclusion

I feel that the use of *ISETL* in my abstract algebra class has improved my students' understanding of many concepts of group theory and their ability for correct mathematical reasoning. In the last few years the number of students resistant to using computers has decreased drastically and students seem more comfortable with learning new programming languages. I am now able to move most of the *ISETL* work outside of the classroom, which leaves plenty of in-class contact hours to discuss the theory. Students armed with a wide variety of examples, and counter-examples, are much more likely to come up with the general arguments required in writing a mathematical proof. Their *ISETL* experience has also taught them to be very careful with mathematical notation.

References

[1] E. Dubinsky and U. Leron, *Learning Abstract Algebra with ISETL*, Springer-Verlag, New York, 1993.

[2] L. Geissinger, *Exploring Small Groups: A Tool for Learning Abstract Algebra*, now only available bundled with [5].

[3] J. Gallian, *Contemporary Abstract Algebra*, 4th ed., Houghton Mifflin, Boston, 1998.

[4] U. Leron and E. Dubinsky, *An Abstract Algebra Story*, The American Mathematical Monthly, **102** (1995), 227–242.

[5] E. Parker, *Laboratory Experiences in Group Theory: A Manual to be used with Exploring Small Groups*, Washington, DC, Mathematical Association of America, 1996.

Ruth I. Berger, Department of Mathematics, Luther College, Decorah IA 52101; `bergerr@luther.edu`.

Teaching Abstract Algebra with *ISETL*

Karin M. Pringle

Abstract. This article describes the author's experiences using the *ISETL* software in abstract algebra courses. The article begins with a brief history of *ISETL*, as well as the model devised for its use. Over time the author has tailored this model to accommodate her own teaching style. Included are several examples of activities that have been given to students. The author comments on how the activities have developed students' understanding and also highlights the benefits and pitfalls of some activities.

1 Introduction

This paper describes my experiences teaching abstract algebra with *Interactive Set Language* (*ISETL*). I have taught the course using only the *ISETL* software and also using both *ISETL* and the *Exploring Small Groups* [3] software. As with any technology, an instructor must first learn the software and then adapt it to her or his own teaching style and philosophy. With each semester that I use technology, I vary computer activities and emphases.

This paper begins by explaining the nature of *ISETL* and how the text *Learning Abstract Algebra with ISETL* [1] by Ed Dubinsky and Uri Leron incorporates this language. Next, I give some results from my first semester using *ISETL*. Following this is a section consisting of examples of exercises using *ISETL* that I have used.

2 What is *ISETL*?

ISETL is a software program that Jack Schwartz and Gary Levin developed in the 1960s and 1970s. In the 1980s and 1990s it has been modified to be used as a teaching tool for discrete mathematics, calculus, and abstract algebra courses. In particular, it has been used with the constructivist model of learning. This model holds that in order to learn, students have to, in some way, make the material their own. That is, a mental construction is made based on experience that can then be accessed to apply to a current situation. The *ISETL* software is designed to aid students in building mental constructions. In the summer of 1995, I attended a conference sponsored by the NSF and given by Ed Dubinsky (a proponent of the constructivist approach) to learn about *ISETL*. Information on *ISETL*, including downloading information, can be found in the appendix and at the web site for this volume.

After attending the NSF conference, I taught an abstract algebra course using *ISETL* and the text [1]. In this text, the first chapter is devoted to learning *ISETL* code. At the same time it reviews or introduces concepts that are used later in the course. In the activities, students must first enter *ISETL* code as it is written in the book and then witness and interpret the output. In this way, all the essential aspects of the code are demonstrated to the students before they must use each piece to write their own code. In fact, all of the *ISETL* components that are needed later in the text are found in Chapter 1. The remaining chapters all follow the ACE cycle: Activities, Class Discussion, Exercises.

2.1 Activities

The activities are written to support group work on a computer. They introduce new concepts for students to experiment with using *ISETL*. Mostly, students see examples of groups and other concepts in this section. They frequently have to read ahead in the text or make conjectures to complete the problems. The emphasis of this stage is for students to make a good effort at solving the problems, not that they solve them correctly. This is the process in which students build mental constructs that will help them understand the theory.

2.2   Class Discussion

The Class Discussion portions of the book contain theorems, examples, and discussion. Proofs of the theorems are most often left to the exercises. The critical point of these discussions is that this material is meant to be covered not by lectures but by directing the students to work together in teams and then discussing their findings with the whole class. The team work involves tasks that lead students to examine the theorems. Due to the enormous time it takes for students to solve tasks, I found myself covering most of the material in lectures interspersed with problems for them to try.

2.3   Exercises

The Exercises portion of the text consists of traditional exercises that one would find in a typical abstract algebra text. I found that it worked best for students to submit both their written exercises and the results from their team activity. Turning in one copy of the computer output per team for the activities promoted cooperation and participation as a team. This naturally led to team participation on the written exercises, which I encouraged. During the first part of the semester, I required students to turn in one copy per team of the written exercises. However, as I felt that more individual responsibility was necessary, I had them each turn in their own written exercises for the remainder of the semester. This system seemed to provide a better balance of group work and individual work. With the small amount of lecturing, and the book's tendency to leave proofs of theorems as exercises, my students did not have many examples of complete proofs. I tried to remedy this with thorough solutions to the exercises. In fact, my only reservation about using technology in this course is that there is less time to develop good proof-writing skills.

Although the authors recommend assigning all the activities, I found that assigning fewer exercises helped alleviate the heavy workload. Maybe because of this, I found that there was too big of a gap between the *ISETL* activities and the homework exercises. Some students got the impression that a proof by example was valid. Another concern was that students used the computer as a black box. For example, in the set of written exercises dealing with homomorphisms, students resisted proving a function was a homomorphism. Instead they conjectured that the *ISETL* function is_hom, that they had coded earlier, would return true.

Each semester that I have taught using computers, some groups of students establish themselves as leaders. Typically these are groups that have one or more students majoring or minoring in computer science or students who are high performers on written assignments and exams. However, even the teams that did well on the computer activities did not always understand and assess the meaning of the activities. They would rush through the activities and turn in a computer printout without the required written explanation. Each time that I have taught this course, I have tended to cut back on the length of the computer activities so that students have more time to better understand each activity. To get the most out of the activities, students need to work at their own pace. Then, when finished, to participate in a discussion about the results of the activity and what theorems are indicated. However, because groups worked at different paces, a class discussion usually was not possible. So, I would discuss the activities with each group as they finished.

3   Results

The first time we used *ISETL*, it was not until the fourth week of classes because of construction of a new wing on the mathematical sciences building. Until that time, I lectured on the first few sections and had students work on activities in teams. By the time the class was settling in to a pattern of computer activities, exercises, and class discussions, more than a month had elapsed. I was already in the habit of doing some lecturing. This is an immediate deviation from the Dubinsky method. As the term progressed, I shortened the computer assignments, provided some of the *ISETL* code, and alternated lecturing with in-class, non-computerized team activities. An issue of concern was the workload. On a survey I gave seven weeks into the term, 66% of the class reported that the workload was too heavy. On a survey at the end of the semester, 77% reported that the course required more work than other similar courses. I think this may be a typical response for an abstract algebra course. I found that my students were not bridging the gap between the ideas in the computer activities and the necessary definitions, theorems, and methods needed in the written assignments. Again, one of the reasons for this may be that they were under time pressure, and so they merely punched in the computer activities to get a printout but spent very little time analyzing the output.

On the exams, students performed best on the items that required them to give examples of groups or subgroups having a specified property. They also did well on exams in which they worked in teams to examine properties of a new set and operation. On the first exam, all teams were successful at proving that a direct product of groups is a group. On the second exam most teams found all the subgroups of the quaternion group and proved that the subgroups are all normal. On the final exam, the class performed best on a proof that used Lagrange's theorem.

4   Examples of Activities using *ISETL*

In subsequent semesters I have used *ISETL* and the text *Contemporary Abstract Algebra* [2], by Joseph Gallian. I switched textbooks mainly because I wanted my students to have a more traditional text with a resource of proofs. I also liked the illustrations of groups and the designs that embody different symmetries. Even though I changed texts, I continued the same format of using one day per week for computer activities with *ISETL*. I also continued having students work together in teams. I have found that communicating ideas and solving problems in teams enhances their understanding of the subject. I modified the activities to reduce the number of examples to test with *ISETL*, but I required students to make conjectures and attempt to prove them with pencil and paper. In this way I shrunk the disparity between the computer activities and the written exercises. I have chosen a few activities to share that investigate areas in which students typically have difficulty but where I felt *ISETL* enhanced their understanding. I have written these in the style of the activities in [1]. I have used these activities in successive semesters.

First, I will present two codes that students have to write. One of the first codes they write is to define the modular groups for any natural number n.

```
Zmod:= func(n);
  if n > 1 then
  return {0..(n - 1)};
  end;
end;

amod:= func(n);
  if n > 1 then
  return func(x, y);
    return (x + y) mod n;
    end;
  end;
end;
```

As an illustration of the first function, the output of Zmod(10) is 0,1,2,3,4,5,6,7,8,9. Similarly, the second function creates a function called amod such that when given three integers n, x, and y, the result is the sum of x and y modulo n. For example, amod(10)(6,7) returns 3. When they are testing different groups for various properties, the procedure name_group standardizes notation. Given a set, subset, and operation, the procedure assigns G, H, o, e, and k to designate the group, subgroup, operation, identity, and inverse function, respectively. They also had to write the code is_group to test whether a set and operation form a group. (See the preceding article by Ruth Berger for more details in writing is_group.)

The first activity deals with permutation groups—recognizing different notations for the elements, investigating the order of elements and cyclic subgroups, and finding inverses of permutations.

Activity 1. Permutation Groups

1. To print S_3, type S3. Identify each element of S_3 with its cycle notation.

2. Print S_4 and calculate the number of elements in S_4. Count the number of permutations of each type according to cycle length and the number of cycles.

3. The composition of permutations is performed by the command `.os`. Run `name_group` on `S4` with `.os`. Find the cyclic subgroup generated by

    (a) `[2,3,4,1]`
    (b) `[2,3,1]`
    (c) `[2,1]`
    (d) `[2,1,4,3]`

4. Find the order of each element in question 3.

5. Use the inverse function to calculate the inverse of each of the following.

    (a) `[2,3,4,1]`
    (b) `[3,1,4,2]`
    (c) `[2,3,1]`
    (d) `[2,1]`
    (e) `[2,1,4,3]`

6. Find the inverses of `[2,3,4,5,1]` and `[2,3,1,5,4]` in S_5.

7. What do you think is the inverse of $[a_2, a_3, a_4, \ldots, a_k, a_1]$? Prove your conjecture.

In item number 3, the link between the two notions of order, that of an element and that of a group is emphasized. Question number 7 is an example of how I use computer activities to lead my students to make their own conjectures that they can then prove. This is a fairly short activity sheet and students typically complete it with little difficulty.

The next activity deals with cosets. One aspect of cosets that is stressed is the normality condition that is necessary for the set of cosets to form a group. In writing traditional proofs and exercises, I rely most on the definition of normality using the invariance of conjugation. When using *ISETL*, a first step to the notion of conjugation is given by checking whether the left cosets and the right cosets are the same. This is a very easy check using *ISETL*. There is a follow-up check to see if the set of left or right cosets form a group in its own right. This activity leads students to the result that these two notions are equivalent. I find that quotient groups and cosets are usually very difficult ideas for students to understand. The following example illustrates an activity involving cosets. In this activity (and later ones also), K_4 represents the Klein-four subgroup $\{[1,2,3,4], [2,1,4,3], [3,4,1,2], [4,3,2,1]\}$, D_4 represents any subgroup of S_4 isomorphic to the dihedral group of order 8, and A_4 represents the alternating subgroup of S_4.

Activity 2. Cosets

1. Input the following code.

    ```
    leftcoset := func(x);
       return {(x .o h) : h in H};
    end;
    rightcoset := func(x);
       return {(h .o x) : h in H};
    end;
    ```

2. For each G and H given below, do these two steps and then answer the following questions: Run `name_group` on G and for each $x \in G$, determine `leftcoset(x)`. (Hint: use `{leftcoset(x) : x in G};`.) How many elements are in each of the subsets? How many distinct subsets are there? Does the set of left cosets equal the set of right cosets?

    (a) $G = \mathbb{Z}_{24}$ and $H = \{0, 4, 8, 12, 16, 20\}$.
    (b) $G = S_3$ and $H = \{[1, 2, 3], [2, 3, 1], [3, 1, 2]\}$.
    (c) $G = S_3$ and $H = \{[1, 2, 3], [2, 1]\}$.
    (d) $G = S_4$ and $H = K_4$.
    (e) $G = S_4$ and $H = D_4$.
    (f) $G = S_4$ and $H = A_4$.

3. In this exercise, we investigate for which a and b we have $aH = bH$. For $G = S_4$ and $H = K_4$, print the set `{leftcoset(x) : x in G}`.

(a) Compare `leftcoset([2,1])` and `leftcoset([4,3])`.

(b) Find all x in G such that `leftcoset(x) = leftcoset([2,1])`.

(c) Compute `[2,1] .o x` for all x in part b.

(d) For $G = \mathbb{Z}_{20}$ and $H = \{0, 4, 8, 12, 16\}$, print the set $\{$`leftcoset(x) : x in G`$\}$.

(e) Repeat part b for the G and H defined in d, using the element 3.

(f) Compute `17 .o x` for all x in part d.

(g) Fill in the blank and prove the completed statement. $aH = bH$ if and only if _____ * _____ is in H.

In this last activity, item number 3 is an attempt to emphasize when two cosets are equal. Even after completing parts 3a–3f, not all students could answer 3g correctly. When a similar question appeared on the final exam, a disappointingly large number of students answered that two cosets are equal only if the representatives are equal.

I became aware of student difficulties that I did not anticipate by listening to students discuss the problems in their teams. Students' difficulty in determining whether the sets `leftcoset(x)` and `rightcoset(x)` were equal was one of these areas. This activity led to the discussion of exactly what was going wrong in the cases where G/H did not form a group and the crucial role of the generalized product and conjugacy.

One of the written exercises that I assigned was to prove that the order of the element xH in G/H divides the order of the element x in G. One activity that leads students to believe this result deals with the order of an element, generating a cyclic subgroup, and the order of the image of an element under the canonical homomorphism. Students wrote code for a function that takes an element and lists all of its powers. This function is called `gen`.

```
gen :=func(g);
  local i,j;
  % .o [g: i in [1..j]: j in [1..#G]];
end;
```

Students also need a procedure to find the product of an element and a set. This is the operation `.oo` that is referenced below. This next activity explores the relationship between the order of an element in G and the order of its image under the canonical map from G to G/H.

Activity 3. The Elements of a Quotient Group

1. For $G = \mathbb{Z}_{24}$ and $H = 6\mathbb{Z}_{24}$, form the quotient group G/H and list all of its elements.

2. Using the functions `gen` and `order`, determine the order of each element `x .oo H` in G/H.

3. Which element is the identity?

4. Find the orders of 6, 3, and 4 in $\mathbb{Z}_{24}$. How do these orders compare with the orders of `6 .oo H`, `3 .oo H`, and `4 .oo H` in G/H?

5. If xH is an element in G/H, compare the order of xH in G/H and the order of x in G. Prove your conjecture.

6. Let $G = D_4$ and $H = \langle r^2 \rangle$, where r^2 corresponds to the $180°$ rotation. Determine whether G/H is a group, and if so, what familiar group is it?

This exercise reinforces the ideas that the subgroup H is the cyclic subgroup generated by 6 and that the order of the subgroup $\langle 6 \rangle$ is the same as the order of the element 6. In number 4, students begin to see critical connections. The elements 6 and $6 + H$ look the same except that $6 + H$ is the identity. For the pairs $(3, 3 + H)$ and $(4, 4 + H)$, students can see the relationship of cosets working and easily compare the orders. The students can also do the same exercise for a symmetric group.

In the fourth semester that I taught abstract algebra with technology, I also used the software *Exploring Small Groups* [3] written by Ladnor Geissinger and the book *Laboratory Experiences in Group Theory* [4] written by Ellen Maycock Parker. I particularly like the lab in [4] focusing on quotient groups because it visually demonstrates when a subgroup is normal and shows that the quotient structure is a group if the subgroup is normal. The combination of using *ISETL* and *Exploring Small Groups* has been very beneficial to my students' understanding of this complex idea.

The fourth activity deals with isomorphisms. So that my students would have time to investigate different maps, I gave them most of the code that they needed. I gave them code to determine if a map is one-to-one or onto and left for them to write the code to determine whether it is a homomorphism. In the process of writing this code, students have to consider (possibly) two different operations and what it really means to preserve the operation.

Activity 4. Isomorphisms

In this activity, you will be working with two groups and a map between them. To access notation for two different groups we use `name_group` and `name_group'`.

1. For a function $f : G \to G'$, complete the following code that tests whether f is an isomorphism.

```
is_iso := func(f);
  return (forall x,y in  G | f(x) = f(y) impl x = y) and
  (forall w in G' | (exists x in G | f(x) = w)) and
  (replace this with the third criterion);
end;
```

2. Use your function `is_iso` to test whether the following functions are isomorphisms. If they are not isomorphisms, indicate a reason why not.

   (a) $p : \mathbb{Z}_{32} \to \mathbb{Z}_8$ defined by $p(x) = 4x \pmod 8$.
       (Example: `name_group(Z(32), a(32)); name_group'(Z(8),a(8)); p:=-x -¿ 4*x mod 8-; is_iso(p);`

   (b) $pp : \mathbb{Z}_8 \to \mathbb{Z}_8$ defined by $pp(x) = 4x \pmod 8$

   (c) $q : \mathbb{Z}_8 \to \mathbb{Z}_8$ defined by $q(x) = x \pmod 8$

   (d) $r : \mathbb{Z}_8 \to \mathbb{Z}_8$ defined by $r(x) = 3x \pmod 8$

   (e) $s : \mathbb{Z}_8 \to \mathbb{Z}_8$ defined by $s(x) = 5x \pmod 8$

   (f) $t : \mathbb{Z}_8 \to \mathbb{Z}_8$ defined by $t(x) = 7x \pmod 8$

   (g) $u : \mathbb{Z}_{10} \to \mathbb{Z}_{10}$ defined by $u(x) = $ `i(x)`. (The element x is mapped to its inverse.)

   (h) $uu : S_3 \to S_3$ defined by $uu(x) = $ `i(x)`

   (i) $v : S_3 \to S_3$ defined by $v(x) = [2, 3, 1]$ `.os` x

   (j) $w : S_3 \to S_3$ defined by $w(x) = [2, 3, 1]$ `.os` x `.os` $[3, 2, 1]$

   (k) $y : S_3 \to S_3$ defined by $y(x) = [2, 1]$ `.os` x `.os` $[2, 1]$.

3. Find all inner automorphisms of S_3. Note that two of them are 2j and 2k. For each map, print the set of elements and their images.
   Example: `w:= -x-¿ [2,3,1] .os x .os [3,2,1]; is_iso(w); {[x, w(x)] - x in G};`
   Are all of the inner automorphisms distinct? What is $\text{Inn}(S_3)$?

In number 2, parts c–f give an example of the fact that the automorphism group of $\mathbb{Z}_n$ is isomorphic to the group U_n. This theorem appears later in the course and I have students refer back to this activity to see that they already have an example. Parts h and i are examples that are not isomorphisms. Parts j and k are examples of inner isomorphisms. Although this worksheet does not have examples of isomorphisms between different groups, it gives the students an idea of which maps are isomorphisms and which maps are not.

Of the above activities, 1 and 3 were easily completed during a 50 minute class. Some teams finished activities 2 and 4 in class, but most either did not finish or were too rushed to gain a satisfactory amount of understanding. While some students are happy to spend time out of class on computer activities, the majority of students, (or possibly just a noisy minority) were not.

5   Conclusion

Technology added a new dimension to my abstract algebra course. One of its advantages was that it gave students with varying learning styles a different way to learn mathematics. *ISETL* is a tool that an instructor can use as much or as little as he or she wishes. Students found that the proofs required in standard exercises were challenging and different from what they had been required to do in other classes. My hope has been that 1) *ISETL* will help my students understand the material, 2) that students will have a wealth of examples with which to refer when starting theoretical problems, and 3) more of my students will be able to grasp the ideas of abstract algebra.

I would not say that *ISETL* is a shortcut to learning. However, I found that this is what my students were expecting. Having to write code was frustrating for my students, and only a few of them were convinced that it helped their learning process. Providing more time for students to write their analysis of computer results and stressing the importance of their analysis would help optimize the role *ISETL* plays in student learning. Overall, the inclusion of computer activities in my abstract algebra courses led to more discussions about the theories presented as well as generating a higher level of enthusiasm among my students.

References

[1] E. Dubinsky and U. Leron, *Learning Abstract Algebra with ISETL*, Springer-Verlag, New York, 1993.

[2] J. Gallian, *Contemporary Abstract Algebra*, 4th ed., Houghton Mifflin, Boston, 1998.

[3] L. Geissinger, *Exploring Small Groups: A Tool for Learning Abstract Algebra*, now only available bundled with [4].

[4] E. Parker, *Laboratory Experiences in Group Theory: A Manual to be used with Exploring Small Groups*, Mathematical Association of America, Washington, DC, 1996.

Karin M. Pringle, Department of Mathematics and Statistics, University of North Carolina at Wilmington, Wilmington, NC 28403; `pringlek@uncwil.edu`.

Using *ISETL* and Cooperative Learning to Teach Abstract Algebra: An Instructor's View

Robert S. Smith

Abstract. In this paper we describe an innovative approach to teaching abstract algebra characterized by collaborative learning and special computer activities. This approach eschews the lecture method, advocates a constructivist approach to learning, embraces cooperative learning, and makes computers and a mathematical programming language, *ISETL*, an integral part of the learning process. In the computer laboratory, teams of students use *ISETL* to construct and internalize mathematical ideas. In the classroom, activities focus on conjectures and shared explorations. In addition to pedagogical considerations, we discuss practical issues related to adoption of these innovative methods.

1 Introduction

This paper discusses an innovative approach to teaching abstract algebra characterized by collaborative learning and special computer activities. In laboratory sessions, groups of students use a mathematical programming language, *ISETL*, to construct and internalize mathematical ideas. Classroom activities focus on shared exploration as well. In addition to pedagogical considerations, we discuss practical issues related to adoption of these innovative methods.

The author has taught abstract algebra using this approach three times, and these classes have been very successful. Based on the author's experience, only about 30% of the students in a traditional abstract algebra course get a good grasp of how to do proofs. In the innovative courses taught by the author, about 50% of the students gained a good grasp of doing proofs. This suggests that students understand concepts in abstract algebra more thoroughly and do proofs better with this innovative approach than with the traditional approach.

2 Background Information

In December 1991, the National Science Foundation funded *Learning Abstract Algebra: A Researched Based Laboratory and Cooperative Learning Approach*. This was a three-year, computer-based project characterized by research in teaching and learning, cooperative learning, and special computer activities using *ISETL*, a mathematical programming language. The principal investigators were Ed Dubinsky, Uri Leron, and Rina Zazkis. The author was a consultant. The grant was implemented at Israel Institute of Technology (Haifa, Israel), Miami University (Oxford, OH), University of Oklahoma (Norman, OK), Purdue University (West Lafayette, IN), and Simon Frazier University (Vancouver, BC).

The approach used at these five institutions was innovative and revolutionary. It eschewed the lecture method, advocated a constructivist approach to learning, embraced cooperative learning, and made computers and a mathematical programming language an integral part of the learning process.

Theoretical considerations and research results on this approach can be found in Dubinsky *et al* [4], Leron & Dubinsky [6], and Zazkis *et al* [10]. Minicourses on the Miami implementation of this approach have been given at the 5th, 6th, 7th, and 8th Annual International Conference on Technology in Collegiate Mathematics [9]. In the summers of 1994 and 1995, workshops on this innovative approach were offered at Purdue.

In this note, we will report on the implementation of this approach at Miami, discuss some practical issues of implementation, and make some observations. It is our hope that this note will be helpful to others interested in teaching abstract algebra using constructivism and cooperative learning.

3   The Course

Miami typically offers three abstract algebra sections per semester. In general, whenever multiple sections of a course are offered and one section is nontraditional, it is important to alert—or perhaps warn—the clientele of its existence. Accordingly, the unusual nature of this approach was noted in the header notes of Miami's Course Planning Guide before each offering. A document was also prepared for students who wanted preregistration information for the course. (A copy of this can be downloaded at the web site for this volume under this article.) In this document and in pre-course conferences with students, we were very candid about the course. We believed that it was as important for students to be aware of the stimulating environment in the course as it was for them to learn of the frustration that could ensue during the first few weeks of the course and of the potentially heavy work load. In its three offerings at Miami, twenty-six students registered for the course and twenty-five students completed it. We believe that the blunt disclosure to potential students was a significant factor in the low drop rate. The grades in these classes were as follows: six A's, ten B's, six C's, and three D's.

3.1   The First Day

On the first day of class, students were given the following documents (each of which can be downloaded from the web site for this volume under the Smith article): *Math 421 Course Requirements & Operating Procedure*; *Course Philosophy for Learning Abstract Algebra with ISETL*; and *Math 421, Section A Student Information Form*. This last item was completed by the students and used in forming the working teams for the course.

3.2   Cooperative Learning

By the second day, small teams of 3 or 4 people were formed and work began. Small team cooperative learning is a powerful pedagogical technique. It is particularly well suited to mathematics for at least the following reasons.

- Small team interaction is designed to help all team members learn concepts and problem-solving strategies.

- Problems in mathematics are ideally suited for team discussion because they have solutions that can be objectively demonstrated. Students can persuade one another by the logic of their arguments.

- Problems in mathematics can often be solved by several different approaches, and students in teams can discuss the merits of different proposed solutions.

- Students of mathematics often learn best what they teach themselves or each other, and small team learning fosters self- and collegial-instruction [1].

Since classes tended to have a variety of majors, a conscious effort was made to mix majors in teams (e.g., one team consisted of three students with majors in mathematics education, psychology, and systems analysis). Each team had at least one team member with some computer experience. While Miami is a residential institution with most of its students living in Oxford, we did have some students who commuted. Fortunately, the commuting students were as faithful to the class and to their teams as the residential students were.

The teams were encouraged to meet as a whole and work collectively. Unfortunately, this did not always occur. Some teams did subdivide into subteams, and when this occurred we encouraged the teams to reassemble and discuss the solutions devised by the subteams. It was emphasized that regardless of whether one was a participant in a solution or not, each member of each team was responsible for each problem and was expected to understand each solution. The teams generally followed this pattern.

We encountered no dysfunctional teams. Had we done so, our position would have been this: What the instructor has joined together, let no one put asunder. It was our view that team members should work out their differences and learn to work together. The resolution of intra-team discord may have required the intervention of the instructor. Only under the most extreme circumstances would we have considered reforming teams to alleviate discord.

In the first two offerings, a conscious effort was made to have mixed-gender teams, and in the third offering there were only single-gender teams. It was our experience that team bonding and mathematical creativity were substantially better in mixed-gender teams.

3.3 Constructivism in the Classroom

Our approach to abstract algebra was primarily designed to provide an environment that fostered students' construction of mathematical ideas. This environment necessitated much more involvement by students than the usual classroom lecture and followed a pattern of

$$\text{Activities} \rightarrow \text{Classroom Interaction} \rightarrow \text{Exercises}.$$

Concepts were first introduced through computer activities that would stimulate the students to engage in mathematical thinking with minimal explanation. (See [4], [5], [6], and [9].) These activities were designed to get students to make appropriate mental constructions. Here is an example of some *ISETL* activities that would help students to construct the notion of the inverse of an element. (For the reader's sake, *ISETL* solutions are given as well.)

- Define $\mathbb{Z}_{12}$ and the function (func) inv on $\mathbb{Z}_{12}$ that produces the additive inverse of $x \in \mathbb{Z}_{12}$.

```
Z12 := {0..11};
inv := func(x);
    if x in Z12 then
    return (choose g in Z12 | (x + g) mod 12 = 0);
    end;
end;
```

- Test your func by computing the additive inverses of 2, 5, 6, 11, and 0.

```
inv(2); inv(5); inv(6); inv(11); inv(0);
```

- Use your func to answer the following questions.

    1. What is the inverse of the inverse of 2? What is the inverse of the inverse of 5?

    ```
    inv(inv(2)); inv(inv(5));
    ```

    2. What is the sum of 6 with its inverse? What is the sum of 11 with its inverse?

    ```
    (6 + inv(6)) mod 12; (11 + inv(11)) mod 12;
    ```

Class time following activities was spent on specific tasks that were intended to lead students to refine their mental constructions of mathematical ideas and come to a more formal understanding of them. Finally, students were given standard exercises to reinforce the mathematical ideas.

3.4 Class Meetings and Student Work Load

While the Miami course was officially a four-credit course scheduled to meet four times a week, we in fact met five times per week. Even though this arrangement obviously entailed more work on the part of students, there were no complaints resulting from this increase in meetings. Students treated the extra meeting as an opportunity to do activities or exercises under the watchful eye of the instructor.

In the first two weeks of the course, we met in the computer lab every day, learning to use *ISETL* by performing routine tasks and by constructing mathematical ideas in this language. (See [4], [5], [6], and [9].) While some of the ideas may have been familiar to the students, many others were completely new and might not become formalized for weeks. Here is an example of some *ISETL* activities that would help students construct the notions of a set of integers modulo n and a coset. These activities were assigned a week or two before the students formally met the concept of a group and more than a chapter before the students formally met the concept of a coset.

Try to predict what would be the result of running the code below. Then run the code and check your prediction. Write out a verbal explanation of what the following code is doing.

```
Z12 := {0..11};
H := {0, 3, 6, 9};
coset := func(x);
    return {(x + h) mod 12 : h in H};
    end;
coset(2); coset(5); coset(3); coset(10); coset(94);
%union{coset(g) : g in Z12};
xmodH := |x -> {(x + h) mod 12 : h in H}|;
xmodH (2); xmodH (5); xmodH (3); xmodH (10); xmodH (94);
```

Students did a tremendous amount of work in the first two weeks of the course. By the end of this period, most students were comfortable with the language. In week three, we began the following schedule: Monday was for activities in the computer lab, Tuesday through Thursday was for classroom interaction, and Friday was for exercises in the computer lab. The twice-weekly computer laboratory meetings certainly did not provide adequate time for the teams to complete their work. The majority of work was in fact performed during other time periods.

3.5 Classroom Interactions

Perhaps the easiest way of describing the classroom interactions is to say that the students were engaged in a modified Socratic environment. Frequently the instructor would give some exposition on a topic that was being studied or summarized the results of students' activities. During class, the instructor would introduce definitions and sometimes state theorems. Occasionally the instructor would even prove some of the theorems. However, more often than not, students would prove the theorems, and students were frequently called upon to formulate their own theorems or counterexamples.

Students were not told how to solve problems. They learned that, before a question was asked, serious thought should be given to the issue. Every effort should be made to resolve the issue before seeking wisdom from the instructor. This process would usually involve significant interaction within the students' team and sometime with other teams. When a question was put to the instructor, it was often greeted with a question. A good example of this is the following. A student approached the instructor and said, "Dr. Smith, I'd like to ask you a question so you can ask me a question and I can find out what I'm really trying to do."

From early on in the course, even before the concept of group was mentioned, students were working with sets that were closed under binary operations, with closed subalgebraic structures, and with group-theoretic concepts. By the time groups were formally introduced, the students had already worked with a variety of examples and group-theoretic concepts. The preferred way of introducing a topic was to have students explore examples relating to the topic before any mention of the topic. In this way, we prepared very fertile ground in which to plant some mathematical seeds.

To illustrate this notion of examples leading to mathematical growth, let us point out that this approach to abstract algebra dictated considerable study of the permutation group S_n and its subgroups. This study was facilitated by students using *ISETL* to construct S_n for specific values of n. This immersion into permutation groups through specialized computer tasks greatly facilitated students' appreciation of and sensitivity to Cayley's Theorem.

4 The Software

Jack Schwartz and his colleagues at Courant Institute are the grandparents of *ISETL*. In the late 1960s, they gave life to *SETL* (SET Language). This language originally ran on mainframe computers fed by punch cards. In the late 1980s, Gary Levin of Clarkson University became the father of *ISETL* [3, 7]. His work was encouraged and inspired by Ed Dubinsky, Nancy Baxter, and Don Muench. Levin ported the language to DOS and Mac, made it interactive, gave it graphics, and introduced many refinements. He named it *ISETL* (Interactive SET Language). Information about obtaining the current version of *ISETL* can be found in the appendix and at the web site for this volume.

ISETL is an efficacious tool in teaching courses other than abstract algebra. For example, *ISETL* is currently being used to teach precalculus, calculus, linear algebra, and discrete mathematics [8].

5 Textbook and *ISETL* Manual

The text, *Learning Abstract Algebra with ISETL* by Dubinsky and Leron, is well thought out and carefully written. Students found it to be quite readable. In addition to the text, we used *Using ISETL 3.0: A Language for Learning Mathematics-DOS Version* [2] by Jennie P. Dautermann. This manual was particularly helpful in the early stages of the course when the students were learning to write code in *ISETL*.

The text begins with a chapter of activities designed to get students up to speed with *ISETL*. This is followed by chapters on groups, subgroups, group homomorphisms, rings, subrings and ideals, ring homomorphisms, and factorization in integral domains. Most of the standard topics in an abstract algebra course are represented in the text, and there is more than enough material to carry the class through a semester.

6 Some Observations

The self- and collegial-instruction dictated by this approach does not allow one to cover the same number of topics as in a traditional abstract algebra course. Our primary concern in teaching the course was that students learn and understand abstract algebra—not to cover a list of prescribed topics. With this in mind, we let the rate at which students completed activities and exercises set the pace for covering topics. Our orientation, combined with the cooperative learning constructivist approach, enabled students to acquire a much deeper understanding of the topics covered. Even with this approach, we were able to get to the middle of the material on rings. However, since students were so well grounded in independent study and group theoretic notions, it would have been easy for them to complete the material on rings on their own.

Based on our experience of being in a department where there were multiple sections of abstract algebra in a given semester, an inevitable problem was student insecurity about the course. Other instructors covered different topics or covered the same topics in a different order. Further, these instructors moved through the material faster. The students in the innovative course learned of these differences and some concluded that they were being shortchanged or inadequately prepared. We found it important to keep up morale and let students know that, at the end of the semester, they would be at least as competent in abstract algebra as their counterparts in the traditional courses.

Another difficulty may emerge in a department that is wed to an abstract algebra syllabus. The probability is very high that the instructor using the innovative approach will not be able to cover a syllabus written for a traditional course. This issue has been known to create havoc in some departments. With this in mind, it would be prudent to have the support of one's chair (and perhaps one's dean in the case of a small college) before embarking on this approach to abstract algebra. It is also important to have the support of faculty who are teaching courses for which abstract algebra is a prerequisite. This approach to teaching abstract algebra can be controversial. Having the support indicated above can go a long way in calming troubled waters.

7 Conclusion

Our observations suggest that students understand concepts in abstract algebra more thoroughly and do proofs better with this innovative approach than with the traditional approach. As we commented earlier, this approach does make it very challenging to cover all of the topics in a traditional abstract algebra course. However, if an instructor's goal in abstract algebra is for students to understand concepts and learn to do proofs rather than to "cover the syllabus," then the constructivist approach, *ISETL*, and cooperative learning can be instrumental in achieving this goal.

References

[1] N. Davidson, *Small-group cooperative learning in mathematics*, Teaching and Learning Mathematics in the 1990's, 1990 Yearbook, The National Council of Teachers of Mathematics, Reston, VA, 52–61.

[2] J. Dautermann, *Using ISETL 3.0: A Language for Learning Mathematics*, West Publishing, St. Paul, MN, 1992.

[3] E. Dubinsky, *ISETL: A Programming Language for Learning Mathematics*, Communications on Pure and Applied Mathematics, **48** (1995), 1–25.

[4] E. Dubinsky, J. Dautermann, U. Leron, and R. Zazkis, *On Learning Fundamental Concepts of Group Theory*, Educational Studies in Mathematics, **27** (1994), 267–305.

[5] E. Dubinsky and U. Leron, *Learning Abstract Algebra with ISETL*, Springer-Verlag, New York, NY, 1994.

[6] U. Leron and E. Dubinsky, *An Abstract Algebra Story*, American Mathematical Monthly, **102** (1995), 227–242.

[7] D. Muench, *ISETL-Interactive Set Language*, Notices of the American Mathematical Society **37** (1990), 276–279.

[8] R. Smith, *ISETL and Cooperative Learning-Vehicles for Learning Calculus*, Proceedings of the Fourth Annual International Conference on Technology in Collegiate Mathematics, Addison-Wesley, Reading, MA, 1993, 399–403.

[9] R. Smith and J. Dautermann, *A New Approach to Teaching Abstract Algebra Using ISETL*, Proceedings of the Fifth Annual International Conference on Technology in Collegiate Mathematics, Addison-Wesley, Reading, MA, 1994, 527–534.

[10] R. Zazkis, E. Dubinsky, and J. Dautermann, *Using Visual and Analytic Strategies: A Study of Students' Understanding of Permutation and Symmetry Groups*, Journal for Research in Mathematics Education, **27** (1996), 435–457.

Robert S. Smith, Department of Mathematics and Statistics, Miami University, Oxford, OH 45056-3414; RSSmith@MUOhio.edu; http://www.muohio.edu/juniorscholars/rssmith.html.

Using *GAP* in an Abstract Algebra Class

Julianne G. Rainbolt

Abstract. This article discusses the use of the software *Groups, Algorithms and Programming* (*GAP*) as a tool in a first undergraduate course in abstract algebra. Specific exercises that the author has used in her classes will be explained and discussed. The software is used to provide students with numerous examples of algebraic objects and to demonstrate the patterns and structure of this area of mathematics. Information about how to acquire and use the software and additional related information is provided. In addition, a web site that contains a large collection of *GAP* exercises for a first course in abstract algebra is described.

1 Introduction

A current trend in pedagogy in mathematics emphasizes an active learning environment. This article suggests ways to use the software *Groups, Algorithms and Programming* (*GAP*) in a first undergraduate course in abstract algebra. The software provides students with a tool they can use to test their understanding of the material. In addition, the computer exercises allow students to discover algebraic concepts, on their own or in groups. The types of computer exercises described below can be organized into two overlapping categories. One category of exercises leads students to formulate conjectures. In most cases these conjectures will actually be well known results. Students will be able to use the data to formulate their own conjectures, which may not be ones the instructor has in mind. Students will, in some cases, suggest conjectures that are false or results that are not well known. With *GAP*, students can test their conjectures and perhaps reformulate them if the conjectures fail. Students become involved in carefully examining data for patterns, which *GAP* produces for this purpose. The other category of exercises is designed to use the software to provide a better understanding of a concept by providing a rich source of examples. Groups and other algebraic structures that are too large or complicated to be thoroughly investigated by students by hand can be easily examined using *GAP*.

2 Using *GAP*

The *GAP* software is free. For downloading information, tutorials, and a help manual, see the appendix or the web site for this volume. If one has never used *GAP*, or has had only a brief exposure to the software, the tutorial provides a quick way to get comfortable with the software and to get a sense of its capabilities. While running the software, one can obtain help by typing ? and then the subject for which help is wanted. For example, `gap> ?union` returns a description of how to find the union of two sets. (The software uses `gap>` as the prompt for input.)

When teaching an undergraduate course in abstract algebra, I require students to learn a very minimal amount of *GAP*. Only the commands needed for a particular exercise are introduced. The software commands are introduced as corresponding algebraic objects are defined or discussed. For example, when groups are defined in class, I also show my students a few ways to use *GAP* to produce groups. After introducing the symmetric group on n elements, students are shown how to create the symmetric group S_4 in *GAP*.

```
gap> G := SymmetricGroup(4);
Sym( [ 1 .. 4 ] )}
```

When subgroups are introduced, the command for creating subgroups is given and used to provide examples. For example, the following commands create subgroups H1 and H2 of $G = S_4$ where H1 is the subgroup generated by the 2-cycle $(1, 2)$ and H2 is the subgroup generated by the 4-cycle $(1, 2, 3, 4)$ and the permutation $(1, 4)(2, 3)$.

```
gap> H1 := Subgroup(G, [(1, 2)]);
Group([ (1,2) ])
gap> H2 := Subgroup(G, [(1,2,3,4), (1,4)(2,3)]);
Group([ (1,2,3,4), (1,4)(2,3) ])
```

Students learn the necessary commands as examples are illustrated. We have designed the computer exercises in such a way that students only need to master at most a few *GAP* commands for each new exercise. The new commands are described during class so that students do not spend time trying to learn the software language. As the semester progresses and the level of abstract algebra material develops, the students' knowledge of the capabilities of *GAP* expands.

3   Illustrations

The remainder of this article illustrates examples of specific lessons that the author has used in her abstract algebra courses. In each example the exercises will be given and discussed. Additional examples can be found at my web page [2]. The computer projects described here and on my web page were written in conjunction with Joseph Gallian. At this web site, one will find a link to our lab manual *Abstract Algebra with GAP* [3]. This manual includes a description of how to introduce the computer exercises to the class and the *GAP* commands needed to work the exercises. Its organization corresponds chapter-by-chapter with Joseph Gallian's textbook, *Contemporary Abstract Algebra* (fifth edition) [1].

The first three examples given below lead to conjectures. In the first example students are led to two conjectures they then prove as homework. This example illustrates using *GAP* to help students discover a result before it is formally stated and proved. The second example also demonstrates using data produced by *GAP* to formulate a conjecture. In addition, this shows how subroutines are used to minimize the amount of time students spend learning software commands. In the third example students are led to formulate an incorrect conjecture. This example illustrates a way to keep students from relying too heavily on the reasoning that seeing enough examples where something is true leads to a theorem. The fourth and final example uses *GAP* to provide illustrations of an algebraic object. Here the `DirectProduct` command is used to demonstrate to students several nonisomorphic groups of order 40.

3.1   Example 1. Producing Conjectures

This example is used when students are first exposed to finite cyclic groups. To do the exercises, students need to learn two new *GAP* commands. The command to create a cyclic group of order n is `CyclicGroup(n);`. For example, `gap> H := CyclicGroup(5);` creates the cyclic group H of order 5. The symbol `H.1` refers to the generator of H. The following student exercises are designed to help them better understand the structure of a cyclic group.

1. Let G be the cyclic group generated by an element a of order n. By the Fundamental Theorem of Cyclic Groups, there is exactly one subgroup of G of order k for each k that divides n. In addition, this theorem asserts that every subgroup of a cyclic group is cyclic. Use *GAP* to find the smallest subgroup of G containing each of the following:

   (a) a^4 and a^6 when $n = 30$.

   (b) a^{10} and a^2 when $n = 30$.

   (c) a^{15} and a^2 when $n = 30$.

2. Repeat exercise 1 for $n = 60$ and $n = 50$.

3. Formulate a conjecture that describes the smallest subgroup of a cyclic group G of order n that contains a^{m_1} and a^{m_2} for any positive integers m_1, m_2, and n, where a is a generator of G and m_1 and m_2 are less than n.

4. Again, let G be the cyclic group generated by an element a of order n. Use *GAP* to find the smallest subgroup of G containing the following. (Type `?Intersection` at the *GAP* prompt to see how to find intersections.)

   (a) $\langle a^4 \rangle \cap \langle a^6 \rangle$ when $n = 30$.

   (b) $\langle a^{10} \rangle \cap \langle a^2 \rangle$ when $n = 30$.

   (c) $\langle a^{15} \rangle \cap \langle a^2 \rangle$ when $n = 30$.

5. Repeat exercise 4 for $n = 60$ and $n = 50$.

6. Formulate a conjecture that describes the smallest subgroup of a cyclic group G of order n that contains $\langle a^{m_1} \rangle \cap \langle a^{m_2} \rangle$ for any positive integers m_1, m_2 and n, where a is a generator of G and m_1 and m_2 are less than n.

In this set of exercises, 1 and 2 are directing students toward a conjecture that they need to formulate in exercise 3. Similarly, exercises 4 and 5 lead to the conjecture of exercise 6. For exercise 1, a student's work may be the following.

```
gap> G := CyclicGroup(30);
gap> a := G.1;
gap> Size(Subgroup(G, [a^4, a^6]));
15

gap> Size(Subgroup(G, [a^10, a^2]));
15

gap> Size(Subgroup(G, [a^15, a^2]));
30
```

Since $\langle a^2 \rangle$ is the cyclic subgroup of G of order 15, students should conclude in part (a) of exercise 1 that the smallest subgroup is $\langle a^2 \rangle$. Similarly, in part (b) the smallest subgroup is also $\langle a^2 \rangle$ and in part (c) the smallest subgroup is $\langle a \rangle = G$. After the student investigates the cases $n = 60$ and $n = 50$, she should be able to arrive at the conjecture: The smallest subgroup of the cyclic group of order n generated by a that contains a^{m_1} and a^{m_2} is $\langle a^{\gcd(m_1, m_2)} \rangle$.

The work for exercise 4 may be the following.

```
gap> G := CyclicGroup(30);
gap> a := G.1;
gap> H1 := Subgroup(G, [a^4]);
gap> H2 := Subgroup(G, [a^6]);
gap> Size(Intersection(Elements(H1), Elements(H2)));
5

gap> H3 := Subgroup(G, [a^10]);
gap> H4 := Subgroup(G, [a^2]);
gap> Size(Intersection(Elements(H3), Elements(H4)));
3

gap> H5 := Subgroup(G, [a^15]);
gap> Size(Intersection(Elements(H5), Elements(H4)));
1
```

A student should conclude in part (a) that the smallest subgroup is $\langle a^6 \rangle = \langle a^{12} \rangle$. After doing all the cases, the student should be able to conjecture that the smallest subgroup of a cyclic group of order n generated by a that contains $\langle a^{m_1} \rangle \cap \langle a^{m_2} \rangle$ is $\langle a^{\text{lcm}(m_1, m_2)} \rangle$.

Students can now be asked to prove their conjectures. Since they have "discovered" the theorems, they will have a better understanding of their content and be more likely to remember them. Often I ask my students to try to prove conjectures that are obtained from exercises in this category. In other cases I prove the conjecture as part of a lecture and in still other cases the conjecture is not proved until a more advanced course. Many of my students find it exciting to discover a result that later appears in the textbook. This experience gives them a sense of accomplishment and an appreciation of the result.

3.2   Example 2. Using Supplied Subroutines

The computer exercises in this example use subroutines written in *GAP*. To download the files containing the code, see the Rainbolt page at the web site for this volume. Place the files in the same directory where the *GAP* software is installed. Many of the exercises that I hand out to my students use previously written subroutines. I do this because I think it is important to minimize the amount of time a student spends learning software commands. The subroutines were created in cases where I thought the commands were too involved or too repetitive. The subroutine files are read into *GAP*, as will be shown in the following example.

The subroutines needed in this example are called *idempotentCount* and *nilpotentCount*. To use the second subroutine, type `gap> Read("nilpotentCount");`. The subroutines take a ring R as input and then output the number of idempotent and nilpotent elements in R. For example, here we find the number of idempotent and nilpotent elements in the ring of integers mod 6.

```
gap> M := Integers mod 6;
(Integers mod 6)
gap> idempotentCount(M);
4
gap> nilpotentCount(M);
1
```

Let $\mathbb{Z}_n$ denote the ring of integers mod n. This ring is often used in examples in a first course in abstract algebra. The following computer exercises investigate the number of nilpotent and idempotent elements in $\mathbb{Z}_n$.

1. Find the number of idempotents in $\mathbb{Z}_n$ for a sufficiently large variety of n. Based on your results answer the following.

   (a) How many idempotents are in $\mathbb{Z}_n$ when n is a power of a prime?

   (b) How many idempotents are in $\mathbb{Z}_n$ when n is equal to the product of two distinct primes? How many idempotents are in $\mathbb{Z}_n$ when n is equal to the product of five distinct primes?

   (c) Formulate a conjecture about the number of idempotents in $\mathbb{Z}_n$.

2. Find the number of nilpotents in $\mathbb{Z}_n$ for a sufficiently large variety of n. Based on your results, answer the same three questions given in exercise 1, replacing *idempotents* with *nilpotents*.

One starts as follows.

```
gap> M := Integers mod 3^2;
(Integers mod 9)
gap> idempotentCount(M);
2
gap> nilpotentCount(M);
3
```

Let $n = p_1^{n_1} p_2^{n_2} \cdots p_k^{n_k}$ where the p_i are distinct primes. Although not always easily conjectured, the goal is for the student to see that $\mathbb{Z}_n$ has 2^k idempotents and $p_1^{n_1-1} p_2^{n_2-1} \cdots p_k^{n_k-1}$ nilpotents, which is then proved in class. In this example, students use *GAP* to produce data that they can use to observe the patterns.

3.3 Example 3. Deceptive Patterns

I have observed that some students believe if they see enough instances where something is true, then it must always be true. The following example was designed to curtail this line of reasoning and can be used after the class is introduced to polynomial rings. This example frequently leads students to deduce an incorrect conjecture.

The command to create a polynomial over the ring R is UnivariatePolynomial(R, [a1,a2, ... , an]); where a1 is the coefficient of the constant term, a2 is the coefficient of the linear term, and so on. The command to factor a polynomial f is Factors(f);. For example, the following shows how to factor the polynomial $x^2 - 1$ over the rationals.

```
gap> f := UnivariatePolynomial(Rationals, [-1, 0, 1]);
-1+x_1^2
gap> Factors(f);
[ -1+x_1, 1+x_1 ]}
```

This exercise investigates the factorization of $x^n - 1$ into its irreducible factors over the rational numbers.

> **Exercise.** Factor $x^n - 1$ into its irreducible factors over the rational numbers for $n = 6, 8, 12, 20$, and 30. On the basis of these data, make a conjecture about the coefficients of the irreducible factors of $x^n - 1$. Test your conjecture for $n = 40, 50$, and 105. (This exercise comes from [1, p. 310].)

Students are led to the conjecture that the coefficients of the factors of $x^n - 1$ are always one of -1, 0, or 1. The conjecture seems fine until $n = 105$, at which point the coefficient 2 appears. This example illustrates to students that something could be true for many cases but still be false in general.

3.4 Example 4. Examples to Promote Understanding

This example does not lead to any specific conjecture but instead provides the student with a source of examples. This exercise is done when the class is first introduced to direct products. At this point in the course, making simple computations with the elements in a direct product and understanding the basic structure of a direct product is the goal. The command in *GAP* to form an external direct product is DirectProduct. For example, first we create the cyclic subgroup $C4$ generated by the 4-cycle $(1, 2, 3, 4)$ in S_4.

```
gap> C4 := Subgroup(SymmetricGroup(4), [(1,2,3,4)]);
Group([ (1,2,3,4) ])}
```

Next, we let S3 denote the symmetric group on three elements and take the direct product of C4 and S3.

```
gap> S3 := SymmetricGroup(3);
Sym( [ 1 .. 3 ] )
gap> D := DirectProduct(S3, C4);
Group([ (1,2,3), (1,2), (4,5,6,7) ])}
```

The last output above describes the group D as the group generated by the permutations $(1, 2, 3)$, $(1, 2)$, and $(4, 5, 6, 7)$. Even though we constructed D as an external direct product, *GAP* represents this group as an internal direct product. Note that D is isomorphic to $S_3 \oplus \mathbb{Z}_4$. The subroutine *orderFrequency* takes a group and its order as input. The output is a list of the number of elements in the group of each possible order. (This file resides on this volume's web page as well, under the Rainbolt article.) This subroutine is a modified version of a subroutine written by Loren Larson.

```
gap> Read("orderFrequency")
gap> orderFrequency(D, 24);
[1,7,2,8,0,2,0,0,0,0,0,4,0,0,0,0,0,0,0,0,0,0,0,0]
```

From the last output, we see that $S_3 \oplus \mathbb{Z}_4$ has one element of order 1, seven elements of order 2, two elements of order 3, eight elements of order 4, two elements of order 6, and four elements of order 12. I assign the following computer exercises on direct products.

1. **By hand** find the number of elements of each order in $D_{10} \oplus \mathbb{Z}_2$. ($D_n$ denotes the dihedral group of order $2n$.)

2. Check your answer to exercise 1 using `orderFrequency`. Note that D_{10} can be entered by writing it as the subgroup of S_{10} generated by $(1,2,3,4,5,6,7,8,9,10)$ and $(1,10)(2,9)(3,8)(4,7)(5,6)$.

3. Use `orderFrequency` to find the number of elements of each order in $D_5 \oplus \mathbb{Z}_4$.

4. Are $D_{10} \oplus \mathbb{Z}_2$ and $D_5 \oplus \mathbb{Z}_4$ isomorphic? Justify your answer.

5. **By hand** find the number of elements of each order in D_{20}.

6. Check your answer to exercise 5 using `orderFrequency`. Is D_{20} isomorphic to either $D_{10} \oplus \mathbb{Z}_2$ or $D_5 \oplus \mathbb{Z}_4$? Justify your answer.

7. Identify five non-isomorphic groups of order 40.

The group $D_{10} \oplus \mathbb{Z}_2$ has 40 elements. Thus, it is small enough that students can compute the orders of all the elements. Since the direct product structure is new to students, many will make mistakes. They get to check their work in exercise 2. Below is the work for exercise 2.

```
gap> C2 := Subgroup(SymmetricGroup(5), [(1,2)]);
Group([ (1,2) ])
gap> D10 := Subgroup(SymmetricGroup(10), [(1,2,3,4,5,6,7,8,9,10),
> (1,10)(2,9)(3,8)(4,7)(5,6)]);
Group([ (1,2,3,4,5,6,7,8,9,10), (1,10)(2,9)(3,8)(4,7)(5,6) ])
gap> G := DirectProduct(D10, C2);
Group([ (1,2,3,4,5,6,7,8,9,10), (1,10)(2,9)(3,8)(4,7)(5,6),(11,12) ])
gap> Read("orderFrequency");
gap> orderFrequency(G, 40);
[ 1, 23, 0, 0, 4, 0, 0, 0, 0, 12, 0, 0, 0, 0, 0, 0, 0, 0, 0, 0, 0, 0,
0, 0, 0, 0, 0, 0, 0, 0, 0, 0, 0, 0, 0, 0, 0, 0, 0, 0 ]
```

As can be seen, $D_{10} \oplus \mathbb{Z}_2$ has 23 elements of order 2. Exercise 3 has the student find the orders of all the elements in another group of order 40. In this case, the `orderFrequency` command provides the following results (where G has been redefined, using commands similar to the above, to be $D_5 \oplus \mathbb{Z}_4$).

```
gap> orderFrequency(G, 40);
[ 1, 11, 0, 12, 4, 0, 0, 0, 0, 4, 0, 0, 0, 0, 0, 0, 0, 0, 0, 8, 0,
0, 0, 0, 0, 0, 0, 0, 0, 0, 0, 0, 0, 0, 0, 0, 0, 0, 0, 0 ]
```

We see that $D_5 \oplus \mathbb{Z}_4$ has eleven elements of order 2. Students can now conclude that $D_{10} \oplus \mathbb{Z}_2$ and $D_5 \oplus \mathbb{Z}_4$ are not isomorphic, since they do not have the same number of elements of each order. In exercise 5 they investigate another group of order 40 and then check their work in exercise 6 (with $G = D_{20}$).

```
gap> orderFrequency(G, 40);
[ 1, 21, 0, 2, 4, 0, 0, 0, 0, 4, 0, 0, 0, 0, 0, 0, 0, 0, 0, 8, 0,
0, 0, 0, 0, 0, 0, 0, 0, 0, 0, 0, 0, 0, 0, 0, 0, 0, 0, 0 ]
```

This output shows that D_{20} has 21 elements of order 2, so it is not isomorphic to either $D_{10} \oplus \mathbb{Z}_2$ or $D_5 \oplus \mathbb{Z}_4$.

Exercises 1–6 show students three non-isomorphic groups of order 40. They are then asked to discover 2 more. If students are having difficulty, they can be given the hint to look for abelian groups. For example, since $\mathbb{Z}_{40}$ is abelian it is not isomorphic to D_{20}, $D_{10} \oplus \mathbb{Z}_2$, or $D_5 \oplus \mathbb{Z}_4$.

It is now fun to have groups of students see how many distinct groups of order 40 they can find. The exercises of this example provide students with an engaging way to practice working with direct products. In addition, the exercises are particularly well suited for group work. The direct products are large enough that students will take advantage of helping each other discover how to compute the orders of the elements.

4 Conclusion

The examples above illustrate the different types of computer exercises that I use in my abstract algebra class. In some cases, the software is used to help students discover results. These conjectures can then be formally proved. Students take pride in discovering a result before it is proved. They are then more likely to remember the theorem. In other cases the software provides numerous examples and helps students understand algebraic concepts. *GAP* provides a resource for examples that would not otherwise be available to students.

The *GAP* software is a sophisticated program used by research mathematicians, and the possibilities for its use are vast. At the same time, the software is accessible to undergraduate students. Students need only be exposed to a small portion of its capabilities. By introducing commands as they are needed and using subroutines, the time students spend learning the software is minimal. Yet students obtain a powerful resource for helping them understand abstract algebra. In addition, working through the computer projects, in groups or individually, creates an active learning environment. *GAP* works well as tool to provide an alternative way to motivate, excite, and actively engage students.

References

[1] J. Gallian, *Contemporary Abstract Algebra*, 5th ed., Houghton Mifflin, Boston, 2002.

[2] J. Rainbolt, personal web page, `http://euler.slu.edu/Dept/Faculty/rainbolt/rainbolt.html`.

[3] J. Rainbolt and J. Gallian, *Abstract Algebra with GAP*, `http://euler.slu.edu/Dept/Faculty/rainbolt/manual.html`.

Julianne G. Rainbolt, Department of Mathematics, Saint Louis University, 221 North Grand, Saint Louis, Missouri 63103; `rainbolt@slu.edu`; `http://www.slu.edu/Dept/Faculty/rainbolt/rainbolt.html`.

Experiments with Finite Linear Groups Using *MATLAB*

George Mackiw

Abstract. Matrix groups over finite fields provide an excellent source of examples and computational problems in introductory presentations of abstract algebra. Students view arithmetic in these groups as natural. In addition, packages such as *MATLAB* may be used to verify group relations, find orders of elements, and count or display elements of subgroups.

Examples, problems, and experiments involving subgroups of $GL(n, p)$, the group of invertible matrices over the integers mod p, are given that should strengthen intuition and understanding, heighten curiosity, and motivate subsequent theory. Nonabelian groups of the type rarely encountered in introductory presentations become available and accessible for student experimentation.

1 Introduction

Good examples are essential in an introductory abstract algebra course since they put flesh on the bones of abstract theory. The cyclic groups $\mathbb{Z}_n$, the dihedral groups D_n, and the symmetric and alternating groups S_n and A_n occur often (for various integers n) as examples in introductory presentations of group theory. The purpose of this note is to argue that another class of groups—the matrix groups over finite fields—can be readily added to this list. These groups provide fertile ground for computer-aided experimentation and can serve as excellent sources of computational problems that can make theoretical discussions more meaningful and interesting. Most students will have had some prior exposure to linear algebra so that calculations in these groups are viewed as natural and are easily understood. Elements of these groups are matrices that students can view, manipulate, and examine concretely. Generators and relations involving them occur not as artificial *a priori* constructs, but as descriptions and generalizations of what occurs in practice.

The well-regarded matrix-based package *MATLAB* [1] makes an excellent companion while introducing and exploring these groups, though any package that allows for manipulation of matrices with entries in the integers mod n could be used. The material that follows has been used in recent years in a two-semester introductory abstract algebra course taught at Loyola College in Maryland, a private four-year liberal arts college. The students are primarily mathematics majors who are in their third or fourth years. These students had already used *MATLAB* in their required sophomore-year linear algebra course, so additional instruction in the software was not necessary.

2 The Groups $GL(n, p)$ and $SL(n, p)$

By $GL(n, p)$, we mean the group of invertible $n \times n$ matrices with entries from the field $\mathbb{Z}_p$, where p is a prime. Alternately, $GL(n, p)$ consists of those matrices that have nonzero determinant whose entries come from the integers mod p. The following sequence of *MATLAB* commands, stored as the function (or M-file) *glnp*, creates an element of the group $GL(n, p)$ by first constructing a random $n \times n$ integer matrix, reducing it mod p, and guaranteeing that the chosen matrix has a nonzero determinant.

85

```
function [A, p] = glnp(n, p)
   randn('seed', sum(100*clock));
   A = round(100*randn(n));
   A = mod(A, p);
   d = mod(det(A), p);
   if d == 0, A = glnp(n, p);
end
```

The command A = glnp(4, 7) might return A =

```
3   4   0   3
1   4   4   6
1   4   5   2
3   5   2   1
```

a 4×4 matrix of determinant 6 with entries in $\mathbb{Z}_7$.

Elements of finite groups have well-defined orders. What, then, is the order of A? The function ord(A,p) is designed to compute orders of matrices in $GL(n,p)$.

```
function e = ord(A, p)
   [n, m] = size(A);
   e = 1;
   X = A;
   while norm(X-eye(n), 1) > 0,
      e = e + 1;
      X = mod(X*A, p);
end
```

Applying this function to the matrix A found above, we see that ord(A,7) yields ans = 336. Students might find it interesting to rapidly construct matrices of "large" order on their own. As another elementary use of ord, recall that if x and y are elements in an arbitrary group, the elements xy and yx, though not necessarily equal, must have the same order. We get a concrete view of this by randomly picking two non-commuting matrices, A and B, in some $GL(n,p)$ and then comparing ord(A*B, p) with ord(B*A, p).

The subgroup of $GL(n,p)$ consisting of those matrices having determinant equal to 1 is $SL(n,p)$. Here are a few suggested exercises for students.

- Show that $SL(n,p)$ is a normal subgroup of $GL(n,p)$.

- Write a *MATLAB* M-file that, given n and p, produces a random matrix in $SL(n,p)$.

- Explain why, for any n, $GL(n,2) = SL(n,2)$.

- List by hand the six matrices in $SL(2,2)$, along with their orders. Note that $SL(2,2)$ is a nonabelian group of order 6 and must then be isomorphic to S_3, the symmetric group on three letters. Are there three objects in sight that $SL(2,2)$ permutes? The nonzero vectors in $\mathbb{Z}_2^2$ are certainly fine candidates.

Exercises and questions such as these, and others included below, were often given as homework assignments.

How does the order of the subgroup $SL(n,p)$ compare with the order of the full group $GL(n,p)$? Students might be challenged to produce an intuitive answer: A matrix in $GL(n,p)$ can have any one of the $p-1$ nonzero entries in $\mathbb{Z}_p$ for its determinant. All things being equal, it would appear that $\frac{1}{p-1}$ of the matrices in $GL(n,p)$ should have determinant equal to 1. More formally, the mapping $\phi : GL(n,p) \to \mathbb{Z}_p^*$, given by $\phi(A) = \det(A) \pmod{p}$, is a surjective homomorphism of groups ($\mathbb{Z}_p^*$ is the multiplicative group of nonzero elements of $\mathbb{Z}_p$), whose kernel is precisely $SL(n,p)$. Since $\mathbb{Z}_p^*$ has order $p-1$, the fundamental theorem of group homomorphisms then asserts that $|SL(n,p)| = |GL(n,p)|/(p-1)$.

We could measure the size of $GL(n,p)$ computationally. For example, the final value of the counter `k` in

```
k = 0;
for a = 0:4, for b = 0:4, for c = 0:4, for d = 0:4,
    v = mod(det([a b ; c d]), 5);
    if v ~= 0, k = k + 1; end
end, end, end, end
```

gives the order of $GL(2,5)$ to be 480.

A general argument to find $|GL(n,p)|$, though, would be more satisfying. We construct one based on the observation that invertible matrices are precisely those square matrices whose columns are linearly independent. The first column of any matrix in $GL(n,p)$ can be any one of the $p^n - 1$ nonzero vectors in $\mathbb{Z}_p^n$; the second column can be any one of the $p^n - p$ vectors that are not in the subspace spanned by the first column; the third column can be chosen from any one of the $(p^n - p^2)$ vectors not in the subspace spanned by the first two columns, and so on. Thus,

$$|GL(n,p)| = (p^n - 1)(p^n - p)(p^n - p^2) \cdots (p^n - p^{n-1}).$$

In particular, $|GL(2,p)| = (p^2 - 1)(p^2 - p)$ and $|SL(2,p)| = |GL(2,p)|/(p-1) = p(p^2 - 1)$.

A well-known corollary to Lagrange's Theorem states that the order of an element in a finite group must divide the order of the group. At the start, using the function `ord(A,p)`, we exhibited a matrix in $GL(4,7)$ of order 336. One can check that 336 is indeed a factor of $(7^4 - 1)(7^4 - 7)(7^4 - 7^2)(7^4 - 7^3)$. Examples similar to this are easy to construct.

3   Constructing Dihedral Matrix Groups

We pose the following problem: Find a subgroup of $GL(2,p)$, for some prime p, that is isomorphic to the dihedral group D_9 (the symmetries of the regular 9-gon). D_9 is a group of order 18 generated by two elements R (rotation) and F (flip) satisfying $R^9 = F^2 = I$ (identity) and $FRF^{-1} = FRF = R^{-1}$.

Turning our attention first to finding a matrix that would play the role of R, we note that, if α is a nonzero element of $\mathbb{Z}_p^*$, then the order of the matrix $\begin{pmatrix} \alpha & 0 \\ 0 & \alpha^{-1} \end{pmatrix}$ is the same as that of the element α. Since the order of α must divide the order of $\mathbb{Z}_p^*$, we need a prime p satisfying $9|(p-1)$. Better yet, we need p with $p \equiv 1 \pmod 9$. The choice $p = 19$ will do, and, with a bit of arithmetic, we find that $\alpha = 4$ is an element of order 9 in $\mathbb{Z}_{19}^*$. Thus, a candidate for R is the matrix $\begin{pmatrix} 4 & 0 \\ 0 & 5 \end{pmatrix}$ in $GL(2,19)$.

Since $R^{-1} = \begin{pmatrix} 5 & 0 \\ 0 & 4 \end{pmatrix}$, we also need to find a matrix, F, of order 2 with

$$F \begin{pmatrix} 4 & 0 \\ 0 & 5 \end{pmatrix} F = \begin{pmatrix} 5 & 0 \\ 0 & 4 \end{pmatrix}.$$

Reminding ourselves of basic properties of matrix arithmetic and the effect that elementary matrices have in matrix multiplication, we might be steered to the choice $F = \begin{pmatrix} 0 & 1 \\ 1 & 0 \end{pmatrix}$. Thus, $\{R^i F^j \mid 0 \le i \le 8, 0 \le j \le 1\}$ forms a subgroup of $GL(2,19)$ isomorphic to D_9. *MATLAB* could be used to multiply group elements, check relations, and compute orders of elements in this group.

Students might ask (or be prompted to ask) questions such as the following.

- In constructing R, why did we not use $\begin{pmatrix} \alpha & 0 \\ 0 & \alpha \end{pmatrix}$?

- Having picked R, is $F = \begin{pmatrix} 0 & 1 \\ 1 & 0 \end{pmatrix}$ the only possible choice?

- Would it have been possible to construct our subgroup in $SL(2,19)$ instead?

4 A Quaternion Subgroup in $SL(2,3)$

The group $SL(2,3)$ has order $24 = 3(3^2 - 1)$ and our attempt to take a closer look at this group begins by generating a table listing the orders of its 24 elements. The *MATLAB* code

```
X = [];
for a = 0:2, for b = 0:2, for c = 0:2, for d = 0:2,
   A = [a b; c d]; v = mod(det(A), 3);
  if v == 1, X = [X ord(A, 3)]; end
end, end, end, end
X
```

produces the display X =

4   6   3   4   6   3   1   3   3   3   3   4   6   3   6   4   2   6   6   6   4   3   6   3   4

of desired orders. Striking, perhaps, is the observation that the group contains only one element of order 2. This must be the matrix $\begin{pmatrix} 2 & 0 \\ 0 & 2 \end{pmatrix}$, which we will denote by -1. Now, the squares of the six elements of order 4 are elements of order 2. These squares must all then equal -1. Do we see shades of the quaternion group of order 8?

We continue and look at an element of order 4, say $A = \begin{pmatrix} 0 & 1 \\ 2 & 0 \end{pmatrix}$. If our hunch regarding a quaternion subgroup is correct, we should then be able to find an element, B, of order 4 with $BAB^{-1} = A^{-1}$. We could run through the list of elements of order 4 and obtain our answer that way, or simply write out the matrix equation

$$\begin{pmatrix} a & b \\ c & d \end{pmatrix}\begin{pmatrix} 0 & 1 \\ 2 & 0 \end{pmatrix} = \begin{pmatrix} 0 & 2 \\ 1 & 0 \end{pmatrix}\begin{pmatrix} a & b \\ c & d \end{pmatrix}$$

and attempt to solve it. One solution is $B = \begin{pmatrix} 1 & 1 \\ 1 & 2 \end{pmatrix}$, a matrix of order 4 in $SL(2,3)$. Since we already have the relations $A^4 = B^4 = 1$ and $A^2 = B^2 = -1$, it follows that $\langle A, B \rangle$, the subgroup of $SL(2,3)$ generated by A and B, is a quaternion group of order 8.

The vector X can also be used to explain why the quaternion subgroup we constructed is the only subgroup of order 8 in $SL(2,3)$ (there are only so many elements of order 4 available) and thus is normal in $SL(2,3)$. A more ambitious assignment is to use X in arguing that $SL(2,3)$ contains no subgroup of order 12, showing that the converse of Lagrange's Theorem is, in general, false. The article [2] contains such an argument, as well as other computational examples involving $SL(2,3)$.

5 Some Subgroups of $SL(3,3)$

The group $SL(3,3)$ has order $3^3(3^3 - 1)(3^2 - 1) = 3^3 \cdot 2^4 \cdot 13$. A subgroup K of order 27 is not difficult to manufacture: The set of matrices of the form $\begin{pmatrix} 1 & * & * \\ 0 & 1 & * \\ 0 & 0 & 1 \end{pmatrix}$ is seen to be closed under multiplication and forms such a subgroup. Experimentation using ord with randomly chosen elements from K produces an interesting observation: Although K is nonabelian, each element in K appears to have order 3.

One way to verify this conjecture is to compute ord(Y,7) for all 27 elements Y in K. But this is only one instance of a broader phenomenon: For any prime $p > 2$, the corresponding subgroup K in $SL(3,p)$ is a nonabelian group of order p^3 in which every element has order p. Heuristic evidence is easy to obtain; can students supply a conceptual argument to explain the general case? What is the corresponding statement for $SL(n,p)$? Hint: Consider the binomial expansion of $(A - I)^p$, where A is in K and I is the identity matrix.

What is the center of K? What is the commutator subgroup? Questions such as these can be pursued by first gathering computational evidence that points the way to the appropriate general answer.

6 Suggested Problems

The following is a short list of additional suggested investigations that are similar in spirit to those pursued above. These problems were used in an independent study setting with a student who had already completed one semester of the abstract algebra sequence.

- Find a subgroup isomorphic to D_{11} in $GL(2, p)$ for some prime p.

- Suppose m is a positive integer dividing $(p - 1)$. Explain how to find a cyclic subgroup of order m in $SL(2, p)$.

- Let Z be the center of the group $SL(2, 3)$. List the elements of Z. To which canonical group is the quotient $SL(2, 3)/Z$ isomorphic?

- Denote by Q the quaternion subgroup of $SL(2, 3)$ we constructed earlier. Note that the matrix $C = \begin{pmatrix} -1 & 0 \\ 0 & 1 \end{pmatrix}$ is an element of $GL(2, 3)$ that is not in $SL(2, 3)$.

  1. Show that both CAC^{-1} and CBC^{-1} are in Q, where A and B are the matrices we used in constructing Q.

  2. Explain why C is an element of the normalizer in $GL(2, 3)$ of the subgroup Q.

  3. Denote by H the subgroup of $GL(2, 3)$ generated by A, B, and C. What is the order of H? (Answer: 16.)

  4. Use *MATLAB* to generate a list of the orders of the individual elements in the subgroup H. Is H isomorphic to the dihedral group D_8? Is H isomorphic to the direct product $Q \times \mathbb{Z}_2$? (Hint: Compare the number of elements of order 2.)

7 Conclusion

Student reaction to using the *MATLAB* package was almost universally positive. A successful tactic in team assignments was to pair those few students whose *MATLAB* experience was less extensive with those who had more experience. In course evaluations, students remarked that exposure to concrete examples and computations enabled them to more fully appreciate subsequent generalizations. In addition to widening the range of available examples, introducing matrix groups over finite fields allowed students to conduct computational experiments that heightened curiosity and led to firmer understanding of abstract theory.

References

[1] *MATLAB*, MathWorks, Inc., Natick, MA, http://www.mathworks.com/.

[2] G. Mackiw, *The Linear Group SL(2,3) as a Source of Examples*, Mathematical Gazette, **81** (1997), March, 64–67.

George Mackiw, Loyola College in Maryland, Baltimore, MD 21210; mackiw@loyola.edu.

Some Uses of *Maple* in the Teaching of Modern Algebra

Kevin Charlwood

Abstract. *Maple* is a valuable supplement to a one-semester course in modern (abstract) algebra to aid in improving the proof-writing skills of our students. The format of our course is lecture and discussion with some emphasis on having students learn a few important concepts by a discovery process enhanced through the use of technology. After a thorough study of examples, students are encouraged to make conjectures and prove their claims. We shall explore how to include *Maple* in the study of modern algebra at an introductory level in the areas of symmetry groups, matrix groups, and groups of complex numbers. We include portions of *Maple* worksheet exercises for each of these three topics; students complete the exercises before the next class meeting to facilitate class discussion. This article is intended for those who want to incorporate some measure of innovation into a typical traditional course.

1   Introduction

At Washburn University, we conduct a one-semester, three-credit course in modern (abstract) algebra every other year. About half of our students are majors in pure mathematics and the other half are majoring in Mathematics for Secondary Teaching. Our emphasis is an axiomatic development of elementary group theory, obtaining results up through the basic isomorphism theorems; elementary ring/field theory is introduced if time permits. As a prerequisite, our students take a "bridge" course in discrete mathematics where they receive their first major exposure in how to read and write proofs. Even with this background, the notion of working with an abstract set together with a binary operation gives many of our students difficulties.

How does *Maple*—or any other Computer Algebra System (CAS)—fit in? Our students get some exposure to *Maple* in our three-semester Calculus sequence, and in Linear Algebra, making use of [1] where needed. In a theoretical course such as Modern Algebra, a CAS has its limitations. While a CAS can perform almost any calculation imaginable, it takes the capacity of human thought to make the necessary theoretical connections (e.g., every subgroup of a cyclic group is cyclic) and to write proofs.

The rationale for using *Maple* in an algebra course was due in part to the difficulty many of our students have with the level of abstraction involved in the types of computations that frequently occur in this course. *Maple* is able to help our students quickly work through enough examples to give them more time to analyze the results of the computations and write about their implications. Virtually all of our mathematics majors come into Modern Algebra with some background in using technology to solve problems. They get to see that *Maple* is more than a "black box" and that it serves as a powerful tool in the exploration of mathematical concepts.

In our Modern Algebra course, students work with concrete examples of symmetry groups, matrix groups, and groups of complex numbers to help them understand the elementary theorems from group theory that we prove in class. We draw on the students' experience in working with *Maple* to do some computations using concrete examples to enable them to draw conclusions and prove theoretical results. The first time the course was taught we used [2] for the course text; the second time we used [3].

There were five students in our most recent course, three of whom were preparing to teach mathematics. Much of the work outlined in each *Maple* worksheet below was accomplished, but with mixed results. One student was concurrently enrolled in Linear Algebra, making the matrix exercises a bit of a struggle. The exercises on complex numbers worked well, due in part to a review of complex numbers and DeMoivre's formula early in the semester. Using *Maple* to multiply permutations during the investigation of orders of permutations was a challenge, since it required writing a single line of *Maple* code to multiply three or more permutations. The author believes this exercise was still worthwhile, as the students had to first fully understand composition of functions on finite sets of letters before the *Maple* code would work properly.

2   Symmetry Groups

We introduce the notion of a group through the symmetries of two and three dimensional geometric figures. First, rigid motions are considered and permutation notation for these motions is introduced. Students are then shown the connection between a sequence of two or more rigid motions taken in succession, and how this translates to the multiplication (composition) of permutations on an appropriate set of n letters. *Maple* helps students understand that multiplication of permutations is the same as composition of the functions on n letters giving rise to each individual permutation. To illustrate using *Maple*, what follows is the requisite code to show that multiplication of cycles is not commutative, but is associative.

```
> # The with(group) command loads the group theory package
> with(group):
> P := [[1, 2, 3]];
> q := [[2, 3]];
> r := [[1, 4]];
> m1 := mulperms(p, q);
> m2 := mulperms(q, p);
> m3 := mulperms(mulperms(p, q), r);
> m4 := mulperms(p, mulperms(q, r));
```

The command `mulperms` performs cycle multiplication, but multiplies from left to right. Most modern algebra texts perform multiplication of permutations from right to left, as one normally does function composition. Thus, students must think about how to write the *Maple* command to get the desired result. We use double-bracket notation for permutations written as cycles, since *Maple* views expressions with single brackets as vectors. This helps to reinforce for students that permutations expressed in cycle form are completely different from row vectors. Students are encouraged to discover how to instruct *Maple* to multiply three or more cycles together. This is not trivial, as students need to think about the order in which they must input the permutations, together with the `mulperms` command, to get the correct output.

Using *Maple* points out the tedium of entering cycles using the proper syntax. For example, after students get comfortable multiplying cycles such as $(1345)(236)(1436)(24)$ by hand, this approach is preferred over using the software. *Maple* is very useful early on in the course, as most of our students struggle with the concept of multiplication of permutations. Several results follow from this work:

1. Students should conjecture that every permutation may be expressed as a product of disjoint cycles, and prove this result.

2. Students should be clear on the noncommutative nature of multiplication of permutations.

3. Students discover the concept of "subgroup generated by" using the `permgroup` command.

4. Students learn about left and right cosets determined by a subgroup of a permutation group.

The group theory package in *Maple* provides an extensive set of commands for working with permutation groups and finitely-presented groups. When Cayley's Theorem is presented later in the course, students begin to see that all one really needs to know about finite groups, from a computational standpoint, is contained within permutation groups.

What follows are some of the details of a worksheet of exercises that our students do during their study of permutation groups. They are encouraged to make use of *Maple* for these exercises. Beyond the worksheets included here, students do *Maple*-related exercises on modular arithmetic, subgroups of a group, cosets, Lagrange's Theorem, and others if time permits.

Worksheet #1

Before our next class meeting, complete the exercises below and give serious thought to the discussion questions that follow. We will discuss these questions in class to motivate the proof of a theorem relating to the concept that you have studied. Make use of *Maple* as much as needed; remember how it multiplies permutations.

Through examples, you will discover what is meant by the *order* of an element in a finite group and how to find it in the case of permutation groups. Design a minimum of four more examples beyond those given in class and see how well you can predict the orders of your permutations.

Discussion questions. Come to our next class meeting prepared to discuss these questions; written responses are to be turned in.

1. What effect do individual cycle lengths have on the order of a permutation?

2. In finding the order of a permutation, does it matter whether or not the cycles are disjoint? Which is preferable?

3. Suppose we have a finite group not presented as a permutation group, but rather as an abstract presentation in terms of generators and relations. How can we use our work on order of elements in permutation groups to find orders of elements in abstractly presented groups? Consider the difference between groups that are abelian and those that are nonabelian.

4. Carefully formulate a theorem that will tell us how to compute the order of a permutation. Be sure to be precise in stating your hypotheses. Give a proof of this theorem.

The first question is designed to get students to look for a relationship between the lengths of the cycles of a given permutation and the order of the permutation. The discovery process is clouded a bit when the students are given permutations that are not expressed in terms of disjoint cycles. The answer to question #2 became apparent after sufficient practice in multiplying permutations. Students were generally able to see that to find an order of a permutation, it is essential to express it as a product of disjoint cycles first and then find the least common multiple of the cycle lengths. Not surprisingly, discussion question #3 was the most challenging. The idea was to get the students thinking about the computational difficulties of finding orders of elements in a nonabelian group.

3 Matrix Groups

Linear Algebra is not a prerequisite to our Modern Algebra course, although most do take it first. Matrix groups and their actions on vector spaces are simply too valuable to ignore, so we give a brief encapsulation of the linear algebra results needed to understand the group structure of special sets of matrices. We concentrate on 2×2 and 3×3 examples since actions on two and three dimensional vector spaces may be easily visualized. In this way, we continue to emphasize the strong connections between algebra and geometry.

First, we load the linear algebra package in *Maple* to allow students to work with various matrix operations. There are three ways to input matrices: by specifying the individual entries, by inputting row vectors, or by using a generating function. The syntax required by *Maple* to express commands reinforces the concepts of vectors and matrices for our students; this helps them better understand these important objects and how to operate with them. The *Maple* code that follows shows that matrix multiplication is not commutative and an example to illustrate the property that the determinant of a product of matrices is the product of determinants (in the 3×3 case).

```
> # First we load the linear algebra package
> with(linalg):
> A := matrix(3, 3, [[2, 1, -3], [0, -1, 1], [-1, 2, 0]]);
> B := matrix(3, 3, [[1, 4, 0], [2, -1, 3], [6, -3, 5]]);
> C := multiply(A, B);
> C1 := multiply(B, A);
> det(A);
> det(B);
> det(C);
```

Students explore different sets of 2×2 and 3×3 matrices under multiplication to decide which form groups and which do not. One of the key concepts needed is that the determinant of a product of matrices is the product of the determinants of the individual matrices. Their work outside of class using *Maple* is tied to work in class where they discuss the following topics:

1. $GL(2)$, $GL(3)$ (general linear groups, including $GL(n)$).

2. $SL(2)$, $SL(3)$ (special linear groups, including $SL(n)$).

3. $U(2)$, $U(3)$ (the real unitary groups, including $U(n)$).

4. $G = \{x \in GL(2) \mid \det(x) = 2^k, k \in \mathbb{Z}\}$.

5. Upper and lower triangular matrices.

6. The noncommutative nature of matrix multiplication.

7. Rotation matrix groups in 2-space and 3-space.

8. Actions of matrix groups on certain sets of vectors.

In the second exercise set that follows, students work with concrete examples involving 2×2 and 3×3 matrices to discover when there is a group structure present.

Worksheet #2

The purpose of these exercises is for you to discover the conditions required for a set of 2×2 or 3×3 matrices to form a group under multiplication. Make use of *Maple* as much as needed; be sure to load the linear algebra package first.

1. Consider the set of 2-vectors

$$T = \{[2, 0], [1, \sqrt{3}], [-1, \sqrt{3}], [-2, 0], [-1, -\sqrt{3}], [1, -\sqrt{3}]\}$$

   and the subgroup H of $SL(2)$ generated by the $60°$ rotation matrix under multiplication. Write out this subgroup as a set. Then, multiply each member of T by two non-identity matrices from the subgroup H and note your results. Sketch a picture in a 2-plane of the set T and also note the action of H on T in your sketch.

2. Choose a minimum of four 3×3 nonsingular matrices and label them. Use *Maple* to multiply them pair-wise and label the products in *Maple*. Remember that matrix multiplication is noncommutative. Now find the determinant of each of the original matrices and of the pair-wise products you have found. What do you notice?

3. Determine if the set of 2×2 matrices having determinant 2^k for $k > 0$ forms a group under multiplication. If it does not, how could you change the condition on the determinant to ensure a group structure on the resulting new set, if at all?

Discussion questions. Come to our next class meeting prepared to discuss these questions; written responses are to be turned in.

1. Refer to #1 above. To what group is H isomorphic, and why? Discuss how to generalize this exercise to 3-space; that is, can you find a set T of 3-vectors and a subgroup H of $SL(3)$ having the same type of action on T as we found in exercise #1?

2. Looking back at #2 and #3 above, what conditions are necessary for a set T of $n \times n$ matrices to form a group under multiplication? You may wish to consider closure and inverses separately.

3. In class we rigorously showed that matrix multiplication is associative for 2×2 and 3×3 matrices. Consider exercise #1 above, and view matrix multiplication of a vector as a linear map on a vector space and give another explanation as to why matrix multiplication is associative.

The first question is designed to remind students that groups may be presented in many different forms; most saw the answer "cyclic group of order 6" immediately from their graph. Discussion question #2 was successful, as most students noted that an essential condition is that the set of determinants must form a subgroup of the real numbers under multiplication. Although we do not get into the topic of groups acting on sets, this exercise provides the necessary geometric motivation for further study of this concept. Students did have some difficulty with the last discussion question in this exercise set. Many had trouble viewing a matrix as a linear function and the multiplication of matrices as composition of linear functions.

4   Groups of Complex Numbers

We introduce the topic of complex numbers early in the term when introducing the notion of a group. Students review the basic operations on complex numbers along with DeMoivre's formula for finding powers and roots of complex numbers. This topic motivates the notion of a cyclic group in our course.

Maple is helpful to work with complex numbers in exact form, especially for products or quotients containing many factors, and for large powers or roots of large order. It helps students understand that certain special sets of complex numbers form groups under multiplication, whereas others do not. They study the roots of unity and related cyclotomic polynomials. Some aspects of this work are greatly simplified by using a CAS. Note that we do not require a special package here. *Maple* uses a capital *I* to denote the imaginary unit. Below is the *Maple* code necessary to illustrate DeMoivre's Theorem in a special case.

```
> # Begin by loading the function "polar" from the function library
> readlib(polar):
> a := polar(2 + 3*I);
> b := (2 + 3*I)^3;
> c := (2 + 3*I)^10;
> abs(a); argument(a);
> abs(b); argument(b);
> abs(c); argument(c);
> # Look closely at the values obtained in the last two computations
```

In this last exercise set, students make use of *Maple* (when needed) to determine if and when subsets of the complex number system form groups under multiplication. In prior class work they have studied the concept of complex nth roots of unity.

Worksheet #3

The point of these exercises is for you to discover the conditions required for a given set of complex numbers to form a group under multiplication. Make use of *Maple* as much as needed.

1. Consider the set of four complex fourth-roots of unity. Give a procedure for finding integral powers of i.

2. Use *Maple* to find exact numerical expressions for the five complex fifth-roots of unity. If needed, factor out $(x - 1)$ from the original polynomial. What special polynomial do we obtain?

3. Let $p = 2$ and $q = 3$. Find the sets of pth and qth roots of unity and graph them together in a 2-plane. (Let x be the real axis and y the imaginary axis.) What is required to form a group of complex numbers that is minimal in the sense that it contains all pth and qth roots of unity? Repeat the exercise for another pair of relatively prime integers p and q. Remember to consider inverses.

4. Let $p = 6$ and $q = 9$ and repeat exercise #3 above. Repeat it yet again for another pair of positive integers p and q that are not relatively prime.

5. Choose a nonzero complex number $a + bi$ having modulus other than 1, and consider the group of complex numbers it generates. Write your $a + bi$ in polar form to help you examine the group it generates. Graph (as best you can) the group you have generated in a 2-plane as a set of points. Note that your group has infinite order. Can you explain why this is so?

Discussion questions. Come to our next class meeting prepared to discuss these questions; written responses are to be turned in.

1. Referring to exercise #3 above, given two relatively prime integers p and q, what is the "smallest" group of complex numbers containing both sets of pth and qth roots of unity? Attempt to prove this result.

2. Using your work in exercises #3 and #4 above, generalize your answer to the first discussion question for any positive pair of integers p and q.

3. Refer to exercise #5 above, considering the polar form of complex numbers. What must be true of the moduli and arguments of the members of a set of complex numbers for it to form a group under multiplication? What condition(s) must hold for it to be a finite group?

This exercise set caused few difficulties. Since discussion questions #1 and #2 are modeled on the notion of forming a new group from cyclic groups of orders p and q, they were able to see that the least common multiple of p and q was the key idea required to obtain a group structure. In question #3, most students readily saw that the condition "modulus of 1" is required for the group to be finite from their work with finite cyclic groups.

A new exercise set under construction will help students discover how the group of linear fractional transformations $f(z) = (az + b)/(cz + d)$ (with $ad - bc \neq 0$) of a complex variable $z = x + yi$ acts on given subsets of the complex plane. They will study the distinction between such transformations that are elliptic, hyperbolic, loxodromic, and parabolic (see [4]). They will also examine the connection between the coefficients of these transformations and 2×2 matrices having these coefficients as entries.

5   Conclusion

The author has taught the Modern Algebra course at Washburn University twice. The innovations discussed here represent a work in progress, and modifications to the exercises and discussion questions will occur when warranted. Students, through all of their successes and struggles with these topics, have so far proven to be the best guide for changes in the approach to instruction.

References

[1] B. Char, J. Devitt, M. Hansen, K. Heal, and K. Rickard, *Maple V Learning Guide*, Springer-Verlag, New York, 1998.

[2] J. Fraleigh, *A First Course in Abstract Algebra*, 5th ed., Addison-Wesley, Reading, MA, 1994.

[3] F. Goodman, *Algebra: Abstract and Concrete*, Prentice-Hall, Upper Saddle River, NJ, 1998.

[4] K. Knopp, *Elements of the Theory of Functions*, Dover Publications, Inc., New York, 1947.

Kevin E. Charlwood, Department of Mathematics and Statistics, Washburn University, Topeka, KS 66621; zzcharlw@washburn.edu.

Using *Mathematica* to Explore Abstract Algebra

Allen C. Hibbard

Abstract. By augmenting *Mathematica* with the *AbstractAlgebra* packages, one can effectively explore many concepts found in an abstract algebra course. These packages encode the functionality to work with groups, rings, and morphisms. Where possible, visualizations of many of the algebraic concepts are also available. Although the *Mathematica* program is required to use the packages, no experience with it is required. After indicating how to obtain, install, and implement *AbstractAlgebra* into a classroom setting, this article gives an overview of the basic structures and capabilities included in the packages. Following this, we illustrate how to create an exploratory environment using *AbstractAlgebra*. We do so by showing how typical questions from abstract algebra can be answered by using these packages.

1 Introduction

In the absence of any capability to work with groups, rings, or morphisms in off-the-shelf *Mathematica*, Ken Levasseur and I began working on filling this gap in 1991. Our goal was to create a collection of *Mathematica* functions that could be used for pedagogical purposes to illustrate and elucidate algebraic concepts. The result was a suite of packages called *AbstractAlgebra* [3]. These packages are freely available from our web site, `http://www.central.edu/eaam/`. We also began writing laboratory notebooks in which we used these packages to construct creative ways of making abstract algebra more understandable for students. These notebooks culminated in *Exploring Abstract Algebra with Mathematica* (*EAAM*) [4], which contains the labs that emerged from our classroom experiences using the packages. This article primarily focuses on illustrating the capabilities of the *AbstractAlgebra* packages and how they can be used but also gives a brief overview of the *EAAM* book.

2 Obtaining and Installing *AbstractAlgebra*

The *AbstractAlgebra* packages can be acquired by two methods: from our web site [3] or from the CD that accompanies [4]. Although the packages are copyrighted (and not in the public domain), they can be downloaded from [3] and used without charge. Alternatively, since [4] contains the software on a CD, includes the printed documentation, and has many ready-to-use labs, many may find this route attractive. In either case, instructions are included that direct the user to place the *AbstractAlgebra* directory into the AddOns/Applications directory of *Mathematica*. After rebuilding the `Help Index`, the packages' extensive documentation is integrated into *Mathematica*'s `Help Browser` in the `Add-ons` section. From there (or the `Master Index`), one can find complete descriptions and illustrated examples of all the functions available in these packages. By either using this article, a startup notebook accompanying the packages, or from additional information found on the web site, one can immediately begin using *Mathematica* to explore abstract algebra.

3 How to Use *AbstractAlgebra*

The *AbstractAlgebra* packages can be used in several distinct ways. The approaches available depend on how one obtained the packages. If one has *EAAM*, then one path is to use the labs included there. Alternatively, if one has only downloaded the packages from our web site, then one needs to build an environment in which to explore algebra. We will consider both of these directions as well as some general principles applicable in either case.

3.1   Using *AbstractAlgebra* Alone

After installing the packages and rebuilding the `Help Index`, one can use the `Help Browser` to learn more about the capabilities of the *AbstractAlgebra* packages. Using the extensive documentation as a guide, one can design exercises or questions for students to pursue. For example, one could ask the question "How does the order of an element in a group compare to the order of the group itself?" By picking a random element from a variety of built-in groups, one can display a table where each row has a group, an element from the group, the order of the element, and the order of the group. With appropriate use of the *AbstractAlgebra* functions `RandomElement`, `GroupoidName`, `OrderOfElement`, and `Order` (and basic *Mathematica* functions such as `Table` and `TableForm`), one can create an environment to effectively explore this question. The details of the code to answer this particular question, as well as many other questions, are the focus of section 5.

   With examples such as these in hand, one can either design laboratory sessions for students as either in-class or out-of-class exercises. Alternatively, if it is possible to display a computer screen in the class, one may choose to use such exercises to guide or enhance a class discussion. In either case, the open-ended environment of the packages is conducive to interactivity. Once an example is considered, it is easy to make slight modifications to consider related problems. For example, using `FormGroupoid` one can consider arbitrary collections of elements under various operations to investigate potential group properties. With the help of the illustrations found in this article, the on-line documentation, and the suite of palettes (particularly the *GroupCalculator*) available to facilitate working with the packages, an instructor is able to design effective environments to explore abstract algebra.

3.2   Using *EAAM*

While we encourage people to use just the packages themselves to explore various questions, this approach is not suitable for everyone. Although knowing the programming language built into *Mathematica* is not necessary to make use of the approach outlined above, having such a background does extend the possibilities of questions and improve the means of pursuing them. Alternatively, another approach is to use *EAAM* ([4]), which is based on the *AbstractAlgebra* packages. This book makes no assumptions about one's *Mathematica* experience; the rudiments are reviewed in an introductory lab. There are 14 group labs and 13 ring labs, all requiring only minimal *Mathematica* skills. The second half of the book consists of a printed version of the on-line documentation.

   Every lab in *EAAM* starts with a set of goals as well as prerequisites. Most labs are independent, though a few assume some experience with a previous lab. Lab questions are posed in the stream of the text where it is natural to ask them, rather than all at the end. The book is intended to *supplement* any main text (whether one uses a groups-first or a rings-first approach). The web site contains correspondences between the labs and many commonly used texts to help instructors sequence the labs. Partial solutions to the labs and suggestions for implementing them are available to instructors. Where possible, the labs are designed to appeal to the visualization of various algebraic ideas. Additionally, they encourage an exploratory environment in which one can make and test conjectures. See the web site [3] for more information (including four sample labs that can be downloaded).

3.3   Common Principles

Whether one uses *AbstractAlgebra* alone or with *EAAM*, there are some common principles to consider. A lab exercise or notebook can be used either as a discovery session in preparation for a class discussion or as a follow-up to a lecture. Which approach to use may depend on the material, the students, and the instructor. I have found it beneficial to have students work together in pairs and jointly submit their set of exercises. Depending on the topic, one may wish to have students work on the exercises in a computer lab for most of the class session, while at other times one may simply help them get started and allow them to finish outside of class. On the other hand, for some more difficult or subtle topics, it may be helpful for the instructor to work through the exercises with the students, helping to guide the discovery and discussion. I have found that using a lab does not need to take time away from course material since often the lab can replace portions of a lecture. As with any computer experience conducted in a classroom setting, it is important that the instructor approaches a lab thoroughly prepared and is able to handle any glitches that may occur. Most importantly, one should always keep in mind that the software is not the focus; it is only intended as a tool to increase the understanding and appreciation of the algebraic concepts at hand.

4 An Overview of *AbstractAlgebra*

There are three basic data structures used in *AbstractAlgebra*: the Groupoid, Ringoid, and Morphoid. A Groupoid is the *Mathematica* data structure used to represent the mathematical concept we call a *groupoid*, which is intended to be a generalization of a group. In this article, a groupoid is a mathematical structure consisting of a set of elements S and a function $\phi : S \times S \to T$, where the set T does not necessarily need to be related to S (though typically it is). Note that this implies we do not necessarily have a binary operation on S in the traditional sense. Similarly, a Ringoid is the data structure representing the mathematical structure of a *ringoid*. We define a ringoid in a fashion similar to a groupoid, except that we now have two operations (again, not necessarily binary in the strictest sense). Finally, a Morphoid is the structure that is used to represent a mathematical function between either two groupoids or two ringoids. Thus, these three data structures are *Mathematica* implementations of generalizations of groups, rings, and morphisms.

To see these data structures in action, we first read in the Master package. (This prepares all the functions to be readily used.)

```
In[1] := Needs["AbstractAlgebra`Master`"]
```

4.1 Groupoids

As indicated above, a Groupoid consists of a set of elements and a "binary" operation. One method of creating a Groupoid is with the FormGroupoid function. For example, consider the groupoid consisting of the set of integers $\{0, 2, 1, 4, 6\}$ under ordinary integer multiplication (indicated by Times). Additionally, we include an option that simply assigns a name to this groupoid.

```
In[2] := G = FormGroupoid[{0, 2, 1, 4, 6}, Times, GroupoidName → "ex. 1"]

Out[2] = Groupoid[{0, 2, 1, 4, 6}, -Operation-]
```

Observe that the output is an object with the label Groupoid, followed by the list of elements and the operation (which in this case is suppressed with the generic name -Operation-). This illustrates the default representation of a groupoid. At other times, the list of elements may be suppressed (if the list is unwieldy) or the operation's name may be revealed. In any case, one can easily extract the elements and the operation of any groupoid by using the following functions. (This may be used if one wishes to work with the elements or reveal the operation.)

```
In[3] := Elements[G]

        Operation[G]

Out[3] = {0, 2, 1, 4, 6}

Out[4] = Times
```

Since a groupoid is meant to have the potential for being a group, we may want to test whether a groupoid satisfies some or all of the group axioms. For example, we can ask whether G has an identity element. (Note that *Mathematica* has a tradition that Boolean-valued functions end with a Q—here we use HasIdentityQ.)

```
In[5] := HasIdentityQ[G]

Out[5] = True
```

In these packages, many functions incorporate additional modes, such as Textual or Visual. Here is the Visual mode for the previous function.

In[6] := **HasIdentityQ[G, Mode → Visual]**

ex. 1		x * y				ex. 1		x * y			

x\y	0	2	1	4	6
0	0	0	0	0	0
2	0	4	2	8	12
1	0	2	1	4	6
4	0	8	4	16	24
6	0	12	6	24	36

x\y	0	2	1	4	6
0	0	0	0	0	0
2	0	4	2	8	12
1	0	2	1	4	6
4	0	8	4	16	24
6	0	12	6	24	36

red: left iden   red: right ide

Out[6] = True

The coloration of this graphical output is meant to indicate that 1 is both the left and right identity. For example, in the picture on the left, the row headed by 1 represents the products of 1 and the elements in the heading row at the top. Since the products match those in the top row, this indicates that 1 is the left identity.

Another group axiom to consider is whether all the elements have inverses. We can test this in the Textual mode. This mode gives the definition of the construct being considered as well as giving specific information for the case at hand. Furthermore, note that in both the Textual and Visual modes, the computation is still given as the actual output.

In[7] := **HasInversesQ[G, Mode → Textual]**

Given a Groupoid G, we say an element g in G has an inverse h if G has an identity e and g * h = h * g = e (where * indicates the operation).

The Groupoid ex.1 contains some elements without inverses. For example, 0 does NOT have an inverse.

Out[7] = False

Closure is another property required to be a group. The Visual mode of ClosedQ follows.

In[8] := **ClosedQ[G, Mode → Visual]**

All the elements marked with yellow are original elements in the set. Those in red are from outside.

ex. 1		x * y			

x\y	0	2	1	4	6
0	0	0	0	0	0
2	0	4	2	8	12
1	0	2	1	4	6
4	0	8	4	16	24
6	0	12	6	24	36

Out[8] = False

Note that the presence of darker elements (red in a colored version) indicate problems, since these elements are not in the original list of elements in the groupoid. Thus, we see this set is not closed under the Times operator.

Finally, the last property required for a group is associativity. Akin to the previous tests, AssociativeQ returns True or False accordingly (after checking all possible triples). In cases where there are many elements, one may alternatively choose to use RandomAssociativeQ.

Instead of testing the axiomatic properties individually, we can also test these together with just one function.

In[9] := **GroupQ[G]**

Out[9] = False

The Cayley table is a device that can reveal a number of useful properties regarding a group. Note below that each element is given a unique color (keyed in the left column and top row). These same colors are then used in the body of the table where the elements occur. (For colored versions of the graphical images, see the web site for this volume; details can be found in the appendix. At the page for this article, there are several ways of viewing all these graphics in color.)

In[10] := **CayleyTable[G, Mode → Visual]**

For each element, a different color is used. The entries in the table corresponding to the elements are then colored and labeled accordingly.

ex. 1   x * y

x\y	0	2	1	4	6
0	0	0	0	0	0
2	0	4	2	8	12
1	0	2	1	4	6
4	0	8	4	16	24
6	0	12	6	24	36

Out[10] = $\{\{0,0,0,0,0\}, \{0,4,2,8,12\}, \{0,2,1,4,6\}, \{0,8,4,16,24\}, \{0,12,6,24,36\}\}$

From the Cayley table itself, one should be able to easily test for closure, an identity, existence of inverses, as well as commutativity. Note also that while the graphic image is useful for conveying information, the actual output is the list of lists that form the rows of the table; this output can then be used by any function that works with lists.

The following illustrates how the right cosets of $\{0, 4\}$ in $\mathbb{Z}_8$ have a well defined operation based on them. Note that this also shows a quotient group. (This can be seen by viewing the pairs of elements $\{0, 4\}$, $\{1, 5\}$, $\{2, 6\}$, and $\{3, 7\}$ each as a single element. We then see that the blocks in the body of the table correspond to these pairs in a well-formed fashion.) Note that specifying Output → Graphics indicates that we want the graphic as the output, instead of the actual cosets (which is the default output). This image was inspired by Ladnor Geissinger's *Exploring Small Groups* [2] software package.

In[11] := **gr1 = RightCosets[Z[8], $\{0, 4\}$, Mode → Visual, Output → Graphics];**

Z[8]   x + y

x\y	0	4	1	5	2	6	3	7
0	0	4	1	5	2	6	3	7
4	4	0	5	1	6	2	7	3
1	1	5	2	6	3	7	4	0
5	5	1	6	2	7	3	0	4
2	2	6	3	7	4	0	5	1
6	6	2	7	3	0	4	1	5
3	3	7	4	0	5	1	6	2
7	7	3	0	4	1	5	2	6

The cosets show up as colored blocks in the first column and top row, as do their products in the body of the table. Now consider the Cayley table for $\mathbb{Z}_4$, created in a similar fashion.

`In[12]:= gr2 = CayleyTable[Z[4], Mode → Visual, Output → Graphics];`

Z[4]   x + y

y / x	0	1	2	3
0	0	1	2	3
1	1	2	3	0
2	2	3	0	1
3	3	0	1	2

Putting the two graphics side-by-side makes it clear to what group the quotient group $\mathbb{Z}_8/\{0, 4\}$ is isomorphic.

`In[13]:= Show[GraphicsArray[{gr1, gr2}]];`

Z[8]   x   y   Z[4]   x   y

y/x	0	4	1	5	2	6	3	7
0	0	4	1	5	2	6	3	7
4	4	0	5	1	6	2	7	3
1	1	5	2	6	3	7	4	0
5	5	1	6	2	7	3	0	4
2	2	6	3	7	4	0	5	1
6	6	2	7	3	0	4	1	5
3	3	7	4	0	5	1	6	2
7	7	3	0	4	1	5	2	6

y / x	0	1	2	3
0	0	1	2	3
1	1	2	3	0
2	2	3	0	1
3	3	0	1	2

In the situation above, the quotient structure naturally arose because the set $\{0, 4\}$ is a normal subgroup of $\mathbb{Z}_8$. The following indicates that the subgroup generated by the permutation $\{3, 2, 1\}$ in S_3 is not normal in S_3.

`In[14]:= NormalQ[H = SubgroupGenerated[Symmetric[3], {3, 2, 1}], Symmetric[3]]`

`Out[14]= False`

Because of this lack of normality, the coset operation is not well defined, as indicated by the failure of having a regular pattern of "square blocks" for products.

`In[15]:= LeftCosets[Symmetric[3], H, Mode → Visual];`

KEY for S[3]: label used → element: {g1 → {3, 2, 1}, g2 → {1, 2, 3}, g3 → {2, 3, 1}, g4 → {1, 3, 2}, g5 → {3, 1, 2}, g6 → {2, 1, 3}}

S[3]   x * y

y / x	g1	g2	g3	g4	g5	g6
g1	g2	g1	g6	g5	g4	g3
g2	g1	g2	g3	g4	g5	g6
g3	g4	g3	g5	g6	g2	g1
g4	g3	g4	g1	g2	g6	g5
g5	g6	g5	g2	g1	g3	g4
g6	g5	g6	g4	g3	g1	g2

In particular, note that when the element g1 = {3, 2, 1} is multiplied by the coset {g3, g4} = {{2, 3, 1}, {1, 3, 2}}, we obtain the coset {g6, g5} = {{2, 1, 3}, {3, 1, 2}}. However, when the element g2 = {1, 2, 3} (which is in the same coset as g1) is multiplied by the coset {g3, g4}, we obtain {g3, g4}. Thus, the coset operation is not well defined, since the product depends on which representative is chosen.

4.2 Ringoids

We now look at some of the constructs available when working with ring-like structures. First we call a function to change our focus, partially implemented to ensure the proper interpretation for $\mathbb{Z}_n$—whether it is a group or a ring.

        In[16] := **SwitchStructureTo[Ring]**

        Out[16] = Ring

FormRingoid works in a fashion analogous to FormGroupoid. The required arguments are the list of elements, the addition operator, and the multiplication operator. Options may be added after these required parameters. Here we consider an extension of the groupoid G that we looked at earlier, using Plus and Times for the addition and multiplication respectively.

        In[17] := **R = FormRingoid[{0, 2, 1, 4, 6}, Plus, Times, FormatElements → True,**
            FormatOperator → False]

        Out[17] = Ringoid[{-Elements-}, Plus, Times]

RingQ is similar to GroupQ; upon the first failure of a required axiom, it returns False. Optionally, one can test each of the ring axioms individually.

        In[18] := **RingQ[R]**

        Out[18] = False

Since there are two operations in a ringoid, each operation requires a separate Cayley table. With ringoids, we use the CayleyTables command instead of CayleyTable; the two tables then appear side-by-side.

In the next example, we form the extension ring of polynomials over the Boolean ring on {1, 2, 3} (whose elements are subsets of the power set of {1, 2, 3} with the Addition being the symmetric difference and the Multiplication being intersection). After doing this, we choose a random polynomial of degree 2 that is monic (that is, the leading coefficient is the unity).

        In[19] := **RandomElement[PolynomialsOver[BooleanRing[3]], 2, Monic → True]**

        Out[19] = {} + {1, 3} x + {1, 2, 3} x^2

While this random polynomial may appear on the surface to look like an ordinary polynomial, it definitely is not. The FullForm of this expression confirms this; the internal form is quite complicated. To verify that this polynomial really is monic, we can ask for the unity of the base ring.

        In[20] := **Unity[BooleanRing[3]]**

        Out[20] = {1, 2, 3}

Next, we consider a random, invertible 3×3 matrix whose elements come from the lattice ring on the divisors of 12 (with the operations LCM/GCD for the addition and GCD for the multiplication).

        In[21] := **RandomMatrix[LatticeRing[12], 3, MatrixType → GL]//MatrixForm**

        Out[21] = $\begin{pmatrix} 4 & 6 & 4 \\ 4 & 12 & 12 \\ 12 & 1 & 12 \end{pmatrix}$

For the last example with rings, we illustrate how we can form the Galois field of order 9. Note that this is given as a quotient ring.

```
In[22]:= GF[9]

Out[22]= Ringoid[{0, x, 2 x, 1, 1+x, 1+2 x, 2, 2+x, 2+2 x}, -Addition-,
         -Multiplication-]
```

4.3 Morphoids

With the standard method of forming a `Morphoid`, the parameter list calls for a function and then either two groupoids or two ringoids, where the specified function uses the first structure as the domain and the second as the codomain. (Technically, the function needs to be, in *Mathematica* parlance, a "pure" function. To express the function $f(x) = 4x^2 + 3x$ as a pure function, one writes `(4#^2 + 3#) &`.) The following creates a morphoid from $\mathbb{Z}_{12}$ to $\mathbb{Z}_6$ using a function that takes an element in the domain and then reduces it modulo 6 to return an element in the codomain.

```
In[23]:= f = FormMorphoid[Mod[#, 6] &, Z[12], Z[6]]

Out[23]= Morphoid[Mod[#1,6] &, -Z[12]-, -Z[6]-]
```

Since our current structure is `Ring`, the domain and codomain are considered as rings. The `MorphismQ` function tests to see if this morphoid is a ring homomorphism, which it is.

```
In[24]:= MorphismQ[f]

Out[24]= True
```

To visually understand why the operations are preserved with a pair of elements, say $(3, 5)$, observe the following illustration. The graphic on the left indicates that the addition is being preserved, while the one on the right shows that the multiplication is being preserved. Follow the arrows and recall that we require $f(a + b) = f(a) + f(b)$ and $f(ab) = f(a)f(b)$ for f to be a ring homomorphism.

```
In[25]:= PreservesQ[f, {3,5}, Mode → Visual]
```

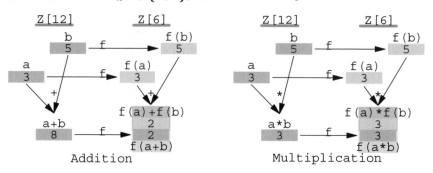

```
Out[25]= True
```

We now switch back to groups to work with group morphisms.

```
In[26]:= SwitchStructureTo[Group];
```

Using the same definition, we now build a *group* homomorphism similar to f defined above.

```
In[27]:= FormMorphoid[Mod[#, 6] &, Z[12], Z[6]]

Out[27]= Morphoid[Mod[#1,6] &, -Z[12]-, -Z[6]-]
```

Note that the output looks identical to that for f. If we would call the `PreservesQ` function again with this morphoid, it would be clear that we are now working with groups.

Sometimes morphisms are more easily defined by matching how we want the elements to correspond. Suppose we want to determine if we can set up a morphism between D_4 and $\mathbb{Z}_8$. The following creates a diagram to help in attempting to construct this.

```
In[28]:= FormMorphoidSetup[D[4],Z[8]];
```

Domain			Codomain
1	1	1	0
Rot	2	2	1
Rot^2	3	3	2
Rot^3	4	4	3
Ref	5	5	4
Rot**Ref	6	6	5
Rot^2**Ref	7	7	6
Rot^3**Ref	8	8	7

Suppose that we want to send the first element in the domain to the first element in the codomain, the second element in the domain to the third element in the codomain, the third to the fifth, and so on. By giving the list of indices of the desired images in the codomain for the preimages whose indices are 1 through 8, we use the FormMorphoid command to create a morphoid. (There is also a newly-designed palette to click one's way to build this morphoid.)

```
In[29]:= h = FormMorphoid[{1,3,5,7,2,4,6,8},D[4],Z[8]]
```

Out[29]= Morphoid[{1→0, Rot→2, Rot^2 →4, Rot^3 →6, Ref→1, Rot**Ref→3, Rot^2**Ref→5, Rot^3**Ref→7}, -D[4]-, -Z[8]-]

Next, we see that this is not a homomorphism on the whole group, since there are some elements that are not colored—read the information printed following the function call.

```
In[30]:= MorphismQ[h, Mode → Visual]
```

The table entry corresponding to the computation a*b in the domain of the morphoid is colored if and only if the pair {a,b} is preserved by the morphoid; i.e., f(a*b) = f(a)*f(b).

KEY for D[4]: label used → element: {g1 → 1, g2 → Rot, g3 → Rot^2, g4 → Rot^3, g5 → Ref, g6 → Rot**Ref, g7 → Rot^2**Ref, g8 → Rot^3**Ref}

D[4]   x * y

X\Y	g1	g2	g3	g4	g5	g6	g7	g8
g1	g1	g2	g3	g4	g5	g6	g7	g8
g2	g2	g3	g4	g1	g6	g7	g8	g5
g3	g3	g4	g1	g2	g7	g8	g5	g6
g4	g4	g1	g2	g3	g8	g5	g6	g7
g5	g5	g8	g7	g6	g1	g4	g3	g2
g6	g6	g5	g8	g7	g2	g1	g4	g3
g7	g7	g6	g5	g8	g3	g2	g1	g4
g8	g8	g7	g6	g5	g4	g3	g2	g1

Out[30]= False

Note, however, that we can see a homomorphism from the rotational subgroup ({g1, g2, g3, g4}) of D_4 to the subgroup {0, 2, 4, 6} of $\mathbb{Z}_8$.

4.4 Summary

This overview is meant to be only a brief introduction of the capabilities of the packages. There are several other sources available to illustrate many of the over 500 functions in *AbstractAlgebra*. Two places to start are the second-half of [4] (containing the documentation for these packages) and our web site [3]. You may also wish to see [5].

5   Questions Answered with *AbstractAlgebra*

This section is meant to illustrate some of the solutions to various questions that can be considered when using the *AbstractAlgebra* packages. These are meant to be sample questions that can act as a springboard to help the reader design his/her own *Mathematica* labs. Some of these are related to questions that are in the labs in [4]. The open-ended nature of the questions is an invitation for further exploration, making conjectures, testing of the conjectures, and finally either refining the conjectures or proving them. Where useful, following a question, a brief description of the *Mathematica* code used to answer it is given in brackets. Note: For each of the following questions, the student should only be expected to answer the *mathematical* question, not provide the *Mathematica* code that provides the framework in which to answer the question. The programming is either done by the instructor (using the *AbstractAlgebra* packages as a foundation) or can be found already available in the labs in [4].

Here is a list of groups that will be used for some of the questions.

```
In[31]:= SomeGroups = {Z[5], Dihedral[4], U[12], Z[18], U[24], Symmetric[3],
    DirectProduct[Z[2], Z[3]], RootsOfUnity[6], Alternating[4], Cyclic[8],
    Klein4}
```

$$Out[31]= \{Groupoid[\{0,1,2,3,4\}, Mod[\#1+\#2,5]\&], Groupoid[\{1, Rot, Rot^2, Rot^3,$$
Ref, Rot**Ref, Rot^2**Ref, Rot^3**Ref\}, -Operation-], Groupoid[\{1, 5, 7, 11\},
Mod[\#1 \#2, 12]\&], Groupoid[\{0, 1, 2, 3, 4, 5, 6, 7, 8, 9, 10, 11, 12, 13, 14,
15, 16, 17\}, Mod[\#1+\#2, 18]\&], Groupoid[\{1, 5, 7, 11, 13, 17, 19, 23\}, Mod[\#1
\#2, 24]\&], Groupoid[\{\{1, 2, 3\}, \{1, 3, 2\}, \{2, 1, 3\}, \{2, 3, 1\}, \{3, 1, 2\},
\{3, 2, 1\}\}, -Operation-], Groupoid[\{\{0, 0\}, \{0, 1\}, \{0, 2\}, \{1, 0\}, \{1, 1\}, \{1,
2\}\}, -Operation-], $Groupoid[\{1, e^{\frac{i\pi}{3}}, e^{\frac{2i\pi}{3}}, -1, e^{-\frac{2i\pi}{3}}, e^{-\frac{i\pi}{3}}\}, e^{i(Arg[\#1]+Arg[\#2])}\&]$,
Groupoid[\{\{1, 2, 3, 4\}, \{1, 3, 4, 2\}, \{1, 4, 2, 3\}, \{2, 1, 4, 3\}, \{2, 3, 1, 4\},
\{2, 4, 3, 1\}, \{3, 1, 2, 4\}, \{3, 2, 4, 1\}, \{3, 4, 1, 2\}, \{4, 1, 3, 2\}, \{4, 2, 1, 3\},
\{4, 3, 2, 1\}\}, -Operation-], $Groupoid[\{1, g, g^2, g^3, g^4, g^5, g^6, g^7\}, -Operation-]$,
Groupoid[\{e, a, b, c\}, -Operation-]\}

Question: How does the order of an element in a group compare to the order of its inverse? [Method: We go through our list of groups just defined, pick a random element from each, calculate its inverse, and then determine the order of both the element and its inverse.]

```
In[32]:= TableForm[Table[G = SomeGroups[[k]]; g = RandomElement[G];
    {GroupoidName[G], g, h = GroupInverse[G, g], OrderOfElement[G, g],
    OrderOfElement[G, h]}, {k, 1, Length[SomeGroups]}], TableHeadings →
    {None, {"group", "g", "g⁻¹", "|g|", "|g⁻¹|\n"}}, TableSpacing →
    {0.5, 3}, TableDepth → 2]
```

Out[32]//TableForm=

group	g	g^{-1}	$\|g\|$	$\|g^{-1}\|$
Z[5]	2	3	5	5
D[4]	Rot	Rot^3	4	4
U[12]	5	5	2	2
Z[18]	5	13	18	18
U[24]	5	5	2	2
S[3]	$\{3, 1, 2\}$	$\{2, 3, 1\}$	3	3
Z[2]×Z[3]	$\{0, 1\}$	$\{0, 2\}$	3	3
RootsOfUnity[6]	$e^{\frac{2i\pi}{3}}$	$e^{-\frac{2i\pi}{3}}$	3	3
A[4]	$\{1, 4, 2, 3\}$	$\{1, 3, 4, 2\}$	3	3
Cyclic[8]	g^5	g^3	8	8
Klein4	c	c	2	2

Question: How does the order of an element in a group compare to the order of the group itself? [Method: Picking a random element from each group, we calculate its order and also the order of the group from which it came.]

```
In[33]:= TableForm[Table[G = SomeGroups[[k]]; g = RandomElement[G];
    {GroupoidName[G], g, OrderOfElement[G, g], Order[G]}, {k, 1,
    Length[SomeGroups]}], TableHeadings → {None, {"group", "g", "|g|",
    "|G|\n"}}, TableSpacing→ {0.5, 3}, TableDepth → 2]
```

```
Out[33]//TableForm=
```

group	g	|g|	|G|
Z[5]	3	5	5
D[4]	Rot^3**Ref	2	8
U[12]	7	2	4
Z[18]	3	6	18
U[24]	23	2	8
S[3]	{3,1,2}	3	6
Z[2]×Z[3]	{0,1}	3	6
RootsOfUnity[6]	$e^{-\frac{2i\pi}{3}}$	3	6
A[4]	{1,3,4,2}	3	12
Cyclic[8]	g^4	2	8
Klein4	a	2	4

Question: Given a random subset of elements from a group, which element(s), if any, need to be added to enlarge this subset to become a subgroup? (Note that the output of the following has been reduced to show the graphic for only one group. The reader can see the complete output by selecting this article at the web site for this volume.) [Method: This code picks a random collection of distinct elements from each group and then visually checks to see if this set is a subgroup.]

```
In[34]:= Table[G = SomeGroups[[k]]; H = RandomElements[G, Random[Integer, {1,
    Order[G]}], Replacement → False]; SubgroupQ[H, G, Mode → Visual], {k, 1,
    Length[SomeGroups]}]
```

All the elements marked with yellow are original elements in the set. Those in red are from outside.

Question: For what values of m and n is $\mathbb{Z}_m \oplus \mathbb{Z}_n$ cyclic? [Method: Allowing m and n each to run from 2 through 7, we simply check whether the direct sum of $\mathbb{Z}_m$ with $\mathbb{Z}_n$ is cyclic.]

```
In[35] := Flatten[Table[G = DirectSum[Z[m], Z[n]]; {m, n, CyclicQ[G]}, {m, 2, 7},
    {n, 2, 7}], 1]//Partition[#, 9] &//Transpose// TableForm[#, TableHeadings →
    {None, {"{m, n, cyclic?}\n"}}, TableSpacing → {0.5, 2}, TableDepth → 2] &
```

```
Out[35]//TableForm=
    {m, n, cyclic?}
```

{2, 2, False}	{3, 5, True}	{5, 2, True}	{6, 5, True}
{2, 3, True}	{3, 6, False}	{5, 3, True}	{6, 6, False}
{2, 4, False}	{3, 7, True}	{5, 4, True}	{6, 7, True}
{2, 5, True}	{4, 2, False}	{5, 5, False}	{7, 2, True}
{2, 6, False}	{4, 3, True}	{5, 6, True}	{7, 3, True}
{2, 7, True}	{4, 4, False}	{5, 7, True}	{7, 4, True}
{3, 2, True}	{4, 5, True}	{6, 2, False}	{7, 5, True}
{3, 3, False}	{4, 6, False}	{6, 3, False}	{7, 6, True}
{3, 4, True}	{4, 7, True}	{6, 4, False}	{7, 7, False}

Question: What is significant about the elements that are generators for $\mathbb{Z}_n$? How are they related to n? Also, how many generators are there, as a function of n? [Method: By looking at the groups $\mathbb{Z}_n$ for $n = 4$ through $n = 10$, we find the generators and then display each of these with their orders.]

```
In[36] := TableForm[Table[G = Z[k]; {Order[G], CyclicGenerators[G]}, {k, 4, 10}],
    TableDepth → 2, TableSpacing → {1, 0.5}, TableHeadings → {None, {"|Z_n|",
    "generators\n"}}]
```

```
Out[36]//TableForm=
    |Z_n|  generators
```

$\lvert\mathbb{Z}_n\rvert$	generators
4	{1, 3}
5	{1, 2, 3, 4}
6	{1, 5}
7	{1, 2, 3, 4, 5, 6}
8	{1, 3, 5, 7}
9	{1, 2, 4, 5, 7, 8}
10	{1, 3, 7, 9}

Question: Given two random cycles, under what circumstances do they commute? Evaluate the following a number of times. [Method: Using the first cycle in the cycle decomposition of two random permutations of length 6, multiply them in the two directions and compare.]

```
In[37] := a = First[ToCycles[RandomPermutation[6]]]

         b = First[ToCycles[RandomPermutation[6]]]

         MultiplyCycles[a, b]

         MultiplyCycles[b, a]
```

```
Out[37] = Cycle[2,5,3,4,6]

Out[38] = Cycle[1,5,4,6,3,2]

Out[39] = {3,1,5,2,6,4}

Out[40] = {5,4,6,3,2,1}
```

Question: Consider the map $\beta : \mathbb{Z}_7 \times \mathbb{Z}_7 \to \mathbb{Z}_7$ defined by $(a, b) \mapsto 2a + 6b$. Is this a homomorphism? What is its kernel? Image? Apply the Fundamental Isomorphism Theorem. [Method: After first defining the direct product and the map, we give output for the following: Is this map a homomorphism? What are the elements in the kernel? What comprises the image? What does the quotient group, formed by modding out by the kernel, look like? Suppose we form the map from this quotient into our image. Is this an isomorphism?]

```
In[41]:= G = DirectProduct[Z[7], Z[7]];
        β = FormMorphoid[Mod[{2,6}.#, 7]&, G, Z[7]];
        MorphismQ[β]
```

Out[43]= True

```
In[44]:= (K = Kernel[β])//Elements
```

Out[44]= {{0,0},{1,2},{2,4}, {3,6},{4,1},{5,3},{6,5}}

```
In[45]:= Img = Image[β]
```

Out[45]= Groupoid[{0,1,2,3,4,5,6}, Mod[#1+#2,7]&]

```
In[46]:= QG = QuotientGroup[G,K]
```

Out[46]= Groupoid[{NS, {0,1}NS, {0,2}NS, {0,3}NS, {0,4}NS, {0,5}NS, {0,6}NS},
    -Operation-]

```
In[47]:= γ = FormMorphoid[β[First[#]]&, QG, Img]
```

Out[47]= Morphoid[β[First[#1]]&, -Z[7]×Z[7]/NS-, -Z[7]-]

```
In[48]:= IsomorphismQ[γ, Cautious → True]
```

Out[48]= True

Question: What can one say about the map from $\mathbb{Z}_{15}$ to $\mathbb{Z}_5$ defined by mapping 2 to 3?

```
In[49]:= VisualizeMorphoid[ZMap[15, 5, 2 → 3]];
```

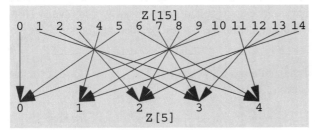

Question: If we take a polynomial of degree 3 and multiply it by a polynomial of degree 2, what will be the degree of the product? [Method: Picking a random polynomial of degree 3 and another of degree 2, compute the degree of the product.]

```
In[50]:= P = PolynomialsOver[ZR[6]];
        TableForm[Table[a = RandomElement[P, 3]; b = RandomElement[P, 2]; {a,
    b, Degree[P, Multiplication[P][a,b]]}, {10}], TableHeadings → {None,
    {"polynomial a", "polynomial b", "deg(a*b)\n"}}, TableSpacing → {0, 3}]
```

Out[51]//TableForm=

polynomial a	polynomial b	deg(a * b)
$1 + 2x + 2x^2 + 3x^3$	$4 + 2x + x^2$	5
$4 + 4x + 4x^2 + 5x^3$	$3x + 3x^2$	5
$5 + 5x^2 + 3x^3$	$2 + 5x + 3x^2$	5
$1 + 3x^2 + x^3$	$5 + 2x^2$	5
$1 + 2x + 5x^2 + 5x^3$	$1 + 2x + 4x^2$	5
$5 + 5x + 4x^2 + 4x^3$	$3 + 2x + 3x^2$	4
$4 + 5x + 4x^2 + 4x^3$	$5 + 2x + 3x^2$	4
$2 + 5x + 4x^2 + 5x^3$	$4 + x + 5x^2$	5
$3 + 4x + 3x^2 + x^3$	$2 + x^2$	5
$3x^2 + 4x^3$	$3x + 2x^2$	5

Question: Is $\mathbb{Z}\left[\sqrt{-6}\right]$ a Unique Factorization Domain? Try evaluating the following lines of code several times. [Method: Picking a random number $x = a + b\sqrt{-6}$, compute its conjugate $y = a - b\sqrt{-6}$ and look at the divisors of the product xy in various ways.]

```
In[52]:= d = -6;

          {a,b} = Table[Random[Integer,{1,5}],{2}];

          x = a + b √d
```

$$\text{Out}[54] = 1 + 3i\sqrt{6}$$

```
In[55]:= y = ZdConjugate[x];

          z = Expand[x y]
```

$$\text{Out}[56] = 55$$

```
In[57]:= ZdDivisors[d, z, DivisorsComplete → True, Combine → Associates]
```

$$\text{Out}[57] = \left\{\{-55,55\}, \{-11,11\}, \{-5,5\}, \{-1,1\}, \{-7-i\sqrt{6},\ 7+i\sqrt{6}\}, \{7-i\sqrt{6},\ -7+i\sqrt{6}\},\right.$$
$$\left.\{-1-3\ i\sqrt{6},\ 1+3\ i\sqrt{6}\}, \{1-3\ i\sqrt{6},\ -1+3\ i\sqrt{6}\}\right\}$$

```
In[58]:= ZdDivisors[d, z, DivisorsComplete → True, Combine → Products]
```

$$\text{Out}[58] = \left\{\{-55,-1\}, \{-11,-5\}, \{1,55\}, \{5,11\}, \{-7-i\sqrt{6}, -7+i\sqrt{6}\}, \{7-i\sqrt{6}, 7+i\sqrt{6}\},\right.$$
$$\left.\{-1-3i\sqrt{6}, -1+3i\sqrt{6}\}, \{1-3i\sqrt{6}, 1+3i\sqrt{6}\}\right\}$$

As can be readily seen, many questions can be explored with functions from the *AbstractAlgebra* packages. For the interested reader, consider the following questions and the suggested hints for answering them. How can one investigate these with *Mathematica*? Actual input and output for these questions can be found at the web site for this volume under this article.

- What can we conclude about the structure of the automorphism group of $\mathbb{Z}_n$? [Hint: Look at the elements of the automorphism group of $\mathbb{Z}_n$ for several values of n.]

- What is the order of an element in a direct product? Is it related to the order of its coordinates? [Hint: Pick random, nonidentity elements from the direct product of a group, e.g., $U_{15} \times \mathbb{Z}_6$, compute the orders of both the elements and its coordinates in the coordinate groups.]

- Determine the structure of the quotient ring $\mathbb{Z}[i]/\langle 2 + 3i \rangle$. [Hint: Use the `QuotientRing` function.]

- Using the Mod p Irreducibility Test (where the degree of the polynomial reduced mod p must be the same as the degree of original), for what p, if any, do we determine whether the polynomial $4x^3 - 5x^2 + x - 8$ is irreducible or not? [Hint: Using the first handful of primes, reduce this polynomial mod the prime p and then consider the images of this reduced polynomial when the domain $\mathbb{Z}_p$ is used.]

- Using $x^4 + x^3 + 1$ as the irreducible polynomial in the Galois field of order 16 (with x as a generator), determine the correspondence between the elements written in multiplicative notation with those given in additive notation. [Hint: Look up the `GF` function.]

6   Closing Remarks

A goal in designing these packages was to increase students' understanding of algebraic concepts using a discovery approach with interactive media and meaningful visualizations. Students using these labs have indicated (in conversations and in class surveys) that this goal has been realized. In particular, some students have expressed their appreciation of the visualizations that helped them make this abstract topic more concrete. Others indicated that the software helped them to more readily come up with conjectures. Since I teach the course using a standard text ([1]), students usually find the inclusion of the labs as an opportunity to strengthen their understanding of the concepts presented in the text.

The *AbstractAlgebra* suite of *Mathematica* packages provides opportunities not currently available with other software applications. The fact that we have individuals from over 45 different countries who have subscribed to our mailing list is a testimony that there is a need for this type of software. The visualization inherent in the *AbstractAlgebra* packages, combined with their extensibility and the exploratory environment of a notebook, makes them an ideal enhancement for teaching and learning abstract algebra.

References

[1] J. Gallian, *Contemporary Abstract Algebra*, 4th ed., Houghton Mifflin, Boston, 1998.

[2] L. Geissinger, *Exploring Small Groups: A Tool for Learning Abstract Algebra*, now only available bundled with [6]

[3] A. Hibbard and K. Levasseur, *AbstractAlgebra* packages, Version 1.0, `http://www.central.edu/eaam/`.

[4] A. Hibbard and K. Levasseur, *Exploring Abstract Algebra with Mathematica*, TELOS/Springer-Verlag, New York, 1999.

[5] K. Levasseur and A. Hibbard, *A Microworld for Learning Abstract Algebra*, Mathematica in Education and Research, **9** (2001), No. 3–4, 4–11.

[6] E. Parker, *Laboratory Experiences in Group Theory: A Manual to be used with Exploring Small Groups*, Mathematical Association of America, Washington, DC, 1996.

Allen C. Hibbard, Department of Mathematics and Computer Science, Central College, Pella, IA 50219; `hibbarda@central.edu`; `http://www.central.edu/~hibbarda`.

Part III

Learning Algebra Through Applications and Problem Solving

The PascGalois Triangle:
A Tool for Visualizing Abstract Algebra

Michael J. Bardzell and Kathleen M. Shannon

Abstract. In this paper we introduce a group-theoretical generalization of Pascal's triangle. We provide triangles constructed by choosing two (possibly the same) group elements, placing one down each side of the triangle and letting group multiplication generate the interior. The triangles presented here have a self-similar quality analogous to that of a fractal. Students can use these structures to visualize certain concepts from abstract algebra including subgroups, quotient groups, and automorphisms. Several student exercises are included for instructors to use in their abstract algebra courses.

1 Introduction

"If only abstract algebra were more visual," students often say. The consensus among many undergraduates is that algebra is the least visual course in the mathematics curriculum. While the Fundamental Theorem of Finite Abelian Groups and Cayley's Theorem are triumphs of classifying abstract structures, there never seem to be many pictures. How does one "see" a quotient group or Lagrange's Theorem? Does a $p-$group "look" any different than a group that is not a $p-$group? The purpose of this paper is to show that abstract algebra can be visual. In fact, many of the concepts just mentioned can be observed in certain pictures. To demonstrate this we turn to a familiar structure.

Pascal's triangle has drawn much attention in combinatorics and number theory. Many patterns can be found in the triangle (for example, see [9]) as well as the mod n version of the triangle. The mod n triangle is constructed by placing 1 down both sides of the triangle and letting addition mod n generate the interior (i.e., an interior entry is found by adding the two entries above it mod n). Pascal's triangles modulo 2^n and modulo a prime are studied in [6] and [10], respectively. A variation using Fibonacci powers is presented in [3]. These papers are written from a combinatorial point of view. Since addition mod n is the group multiplication for the cyclic group $\mathbb{Z}_n$ of order n, we decided to approach Pascal's triangle from a group theoretic point of view. We were curious about the consequences of using noncyclic and nonabelian groups. In this paper we provide new triangles constructed by choosing two (possibly the same) group elements, placing one down each side of the triangle and letting group multiplication generate the interior as in the standard Pascal's triangle. If we place $a \in G$ down the left side and $b \in G$ down the right, then the resulting triangle is denoted (P_G, a, b). As an example, Table 1 contains the first

```
                    1
                 1     2
              1     0     2
           1     1     2     2
        1     2     0     1     2
     1     0     2     1     0     2
  1     1     2     0     1     2     2
1     2     0     2     1     0     1     2
```

Table 1: Eight rows of $(P_{\mathbb{Z}_3}, 1, 2)$.

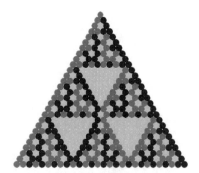

Figure 1: left: $(P_{\mathbb{Z}_3}, 1, 2)$ with 28 rows;   right: $(P_{\mathbb{Z}_3}, 1, 2)$ with 82 rows

eight rows of $(P_{\mathbb{Z}_3}, 1, 2)$. We arbitrarily put a in the top entry of the triangle. When a and b are clear from context, we simply write P_G. For example, throughout this paper $P_{\mathbb{Z}_n}$ will mean $(P_{\mathbb{Z}_n}, 1, 1)$, i.e., Pascal's triangle mod n.

We call (P_G, a, b) a *PascGalois Triangle*. It was Galois who first introduced the term *group* in a technical sense. He is also credited with laying the foundations of group theory (see [5] and [12]). Since the "cal" in Pascal and "Gal" in Galois are phonetically similar, it seemed appropriate to splice the words together as PascGalois to describe these new triangles.

Some interesting patterns emerge when one constructs P_G and assigns a color to each group element. Structures reminiscent of fractals and other self-similar objects become apparent. The main difference is that the self-similar structure is observed by zooming out rather than zooming in as one does with a fractal. Although the pictures in this paper are black and white, the color versions are even more striking. The interested reader can view them from the web site for this volume under this article; see the appendix for details. For example, consider $(P_{\mathbb{Z}_3}, 1, 2)$ shown in Figure 1 (which was numerically presented in Table 1). This is similar to Pascal's triangle mod 3. Note the self-similar structure as we compare the left triangle with 28 rows and the right one with 82 rows. A triangle seems to replicate itself every time we essentially triple the number of rows (i.e., look at the first $3^q + 1$ rows for $q \geq 3$).

In this paper we provide additional examples of this self-replicating phenomenon and relate it to group structures. In particular, we relate the growth and internal symmetries of P_G to subgroups, quotients, and automorphisms of G. We also provide student exercises to facilitate visualizing these concepts. We hope instructors will find these exercises valuable supplements to traditional abstract algebra assignments. The exercises are deliberately loosely posed to encourage student exploration. While there are certain concepts we hope students will obtain from these exercises, student experimentation may lead to other observations or conjectures we have not considered.

PascalGT, a program that will draw PascGalois triangles and hexagons, can be downloaded from our web site [1]. (There is also a link at the web site for this volume under this article.) We also have a handout at our site called the *Triangle Coloring Sheet*. It contains a triangle with 25 rows of blank entries that can be filled in by hand with crayons or markers. However, for numerous examples 25 rows do not reveal many of the patterns. Much more can be seen using the *PascalGT* package. For those who do not wish to create their own images, there is a library of images to accompany this paper. This library can be found at [11] and at the web site for this volume (under this article—see the appendix).

Exercise 1.1 *Using PascalGT or the Triangle Coloring Sheet, construct $P_{\mathbb{Z}_2}$, $P_{\mathbb{Z}_5}$, and $P_{\mathbb{Z}_{10}}$. (Students may also use the image library cited above, if necessary). What patterns and properties do you notice in these pictures? (The $P_{\mathbb{Z}_2}$ and $P_{\mathbb{Z}_5}$ triangles can best be understood by considering the order of the respective group and group elements. The $P_{\mathbb{Z}_{10}}$ triangle should be examined with the subgroup lattice of $\mathbb{Z}_{10}$ in mind).*

Instructors may want to have students study the $P_{\mathbb{Z}_2}$ and $P_{\mathbb{Z}_5}$ triangles first. Students should see a relationship between the number of rows required for self-similarity and the order of the group and group elements. Then, have students examine the $P_{\mathbb{Z}_{10}}$ triangle. They should "see" the $P_{\mathbb{Z}_2}$ and $P_{\mathbb{Z}_5}$ triangles inside of it. In Exercise 5.1 the students learn why this occurs.

2   The Abelian Case

Let G be an abelian group and $a, b \in G$. If we place a down the left side of the triangle and b down the right, then the group multiplication induces the general results shown in Table 2. Let V denote the central vertical axis (row)   (PascGalois triangle)

$$
\begin{array}{c}
0 \qquad\qquad\qquad\qquad a \\[4pt]
1 \qquad\qquad\qquad a \qquad\quad b \\[4pt]
2 \qquad\qquad a \qquad ab \qquad b \\[4pt]
3 \qquad\quad a \quad a^2b \quad ab^2 \qquad b \\[4pt]
4 \quad\; a \quad a^3b \quad a^3b^3 \quad ab^3 \qquad b \\[4pt]
\vdots \\[4pt]
(n+1) \quad a \quad a^{\binom{n}{1}}b^{\binom{n}{0}} \quad a^{\binom{n}{2}}b^{\binom{n}{1}} \quad \cdots \quad a^{\binom{n}{n-1}}b^{\binom{n}{n-2}} \quad a^{\binom{n}{n}}b^{\binom{n}{n-1}} \quad b
\end{array}
$$

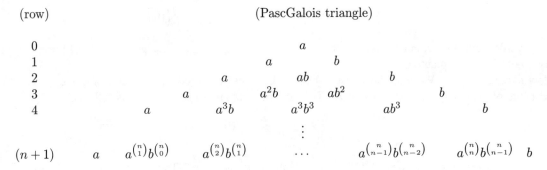

Table 2: General PascGalois triangle.

that bisects P_G. If we reflect P_G about V then we have the correspondence given in Table 3. Using the fact that $\binom{n}{k} = \binom{n}{n-k}$, we have $a^{\binom{n}{n-j+1}} = a^{\binom{n}{j-1}}$. So in general $a^{\binom{n}{j}}b^{\binom{n}{j-1}} \longleftrightarrow a^{\binom{n}{j-1}}b^{\binom{n}{j}}$. In other words, a reflection about V just interchanges the exponents of a and b.

$$
\begin{array}{ccc}
a & \longleftrightarrow & b \\[4pt]
a^{\binom{n}{1}}b^{\binom{n}{0}} & \longleftrightarrow & a^{\binom{n}{n}}b^{\binom{n}{n-1}} \\[4pt]
a^{\binom{n}{2}}b^{\binom{n}{1}} & \longleftrightarrow & a^{\binom{n}{n-1}}b^{\binom{n}{n-2}} \\[4pt]
& \vdots & \\[4pt]
a^{\binom{n}{j}}b^{\binom{n}{j-1}} & \longleftrightarrow & a^{\binom{n}{n-j+1}}b^{\binom{n}{j}}
\end{array}
$$

Table 3: Correspondences from a central reflection.

Exercise 2.1 *Using the PascalGT program or the Triangle Coloring Sheet, construct (P_G, a, b) for various values of m and n where $G = \mathbb{Z}_n \times \mathbb{Z}_m$, $a = (0, 1)$, and $b = (1, 0)$. What happens when you reflect P_G about the central axis V? Does this reflection ever induce a well-defined function $f : G \to G$? If it does, is there anything special about this function?*

As an example let us consider the Klein-4 group, $\mathbb{Z}_2 \times \mathbb{Z}_2$, using $a = (0, 1)$ and $b = (1, 0)$. We obtain the pictures in Figure 2. We know that a reflection about V for $P_{\mathbb{Z}_n}$ moves each color onto itself. That is, reflection induces the identity map $1_{\mathbb{Z}_n}$. In the Klein-4 case, we see that a reflection about V induces another map $\phi : \mathbb{Z}_2 \times \mathbb{Z}_2 \longrightarrow \mathbb{Z}_2 \times \mathbb{Z}_2$, different from the identity map. In fact, ϕ is a nontrivial group automorphism. It is the automorphism that takes $(0, 1)$ to $(1, 0)$ and $(1, 0)$ to $(0, 1)$. This is a consequence of the following result that solves Exercise 2.1.

Theorem 1 *Reflecting $(P_{\mathbb{Z}_n \times \mathbb{Z}_m}, (0, 1), (1, 0))$ about V defines a nontrivial group automorphism if and only if $n = m$.*

Proof.   First assume $n = m$. Throughout this proof, $\overline{\alpha}$ means $\alpha \bmod n$. We have already seen that $a^{\binom{n}{j}}b^{\binom{n}{j-1}} \leftrightarrow a^{\binom{n}{j-1}}b^{\binom{n}{j}}$. However, in $\mathbb{Z}_n \times \mathbb{Z}_n$ $a^{\binom{n}{k}} = \left(0, \overline{\binom{n}{k}}\right)$ and $b^{\binom{n}{k}} = \left(\overline{\binom{n}{k}}, 0\right)$. So reflection about V yields

Figure 2: left: $P_{\mathbb{Z}_2 \times \mathbb{Z}_2}$ with 33 rows;   right: $P_{\mathbb{Z}_2 \times \mathbb{Z}_2}$ with 65 rows

$$\left(0, \overline{\tbinom{n}{j}}\right)\left(\overline{\tbinom{n}{j-1}}, 0\right) = \left(\overline{\tbinom{n}{j-1}}, \overline{\tbinom{n}{j}}\right) \longleftrightarrow \left(\overline{\tbinom{n}{j}}, \overline{\tbinom{n}{j-1}}\right) = \left(0, \overline{\tbinom{n}{j-1}}\right)\left(\overline{\tbinom{n}{j}}, 0\right).$$

In other words, reflection about V induces the map $f : \mathbb{Z}_n \times \mathbb{Z}_n \to \mathbb{Z}_n \times \mathbb{Z}_n$ given by $f(r, s) = (s, r)$. This is clearly a nontrivial group automorphism.

Now assume $n \neq m$. We will show that reflection does not even produce a function from $\mathbb{Z}_n \times \mathbb{Z}_m$ to itself. Without loss of generality assume that $n < m$. Then the nth and $(n+1)$st rows are as follows:

$$
\begin{array}{ccccccc}
 & (0,1) & & (1, n-1) & \cdots & (n-1, 1) & (1, 0) \\
(0,1) & & (1, n) & & \cdots & (0, 1) & (1, 0)
\end{array}
$$

When we reflect the $(n+1)$st row, $(0,1)$ is sent to both $(1,0)$ and $(1,n)$. The result therefore follows. ∎

3 Some Nonabelian Examples

In this section we construct PascGalois triangles using nonabelian groups. The first natural candidate is S_3, the permutation group on three letters. If we place the 2-cycle (1 2) down the left side of the triangle and the 3-cycle (1 2 3) down the right, then using left-to-right multiplication we have the result in Figure 3. Note that there does not appear to be a discernible pattern here. However, if we go further down the triangle something interesting does occur, as shown in Figure 4. Here a definite pattern is emerging. In fact, we can understand this pattern using a quotient

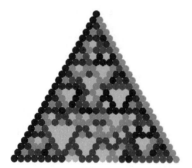

Figure 3: P_{S_3} with 25 rows

Figure 4: P_{S_3} with 125 rows

group. Consider Pascal's triangle mod 2 (the middle picture in Figure 5). There are numerous downward pointing monochromatic sub-triangles in this image. In P_{S_3} (the left picture in Figure 5), there are also downward pointing sub-triangles. However, they are not monochromatic. Close inspection reveals that each of the sub-triangles contains the same three group elements. These elements are the even permutations, i.e., elements of the normal subgroup A_3. The similarity between the P_{S_3} and $P_{\mathbb{Z}_2}$ is due to the fact that $S_3/A_3 \cong \mathbb{Z}_2$. If we color all the even permutations one color and all the odd permutations a second color, then we should obtain the $P_{\mathbb{Z}_2}$ triangle. Instructors should have their students study Figure 5 carefully—it should help them visualize the quotient identification $S_3/A_3 \cong \mathbb{Z}_2$ and the corresponding cosets. The cautious reader, however, may notice a slight difference. In $P_{\mathbb{Z}_2}$, ones appear down both sides. However, in P_{S_3} there is a two cycle down the left side and a three cycle down the right. So when

Figure 5: left: P_{S_3} with 65 rows; middle: $(P_{\mathbb{Z}_2}, 1, 1)$ with 65 rows; right: $(P_{\mathbb{Z}_2}, 1, 0)$ with 65 rows

we do the quotient identification making all even permutations 0 and all odd permutations 1, a 1 appears down the left side of the triangle and a 0 down the right. To remedy this difference we try an alternate construction of the $\mathbb{Z}_2$ triangle. Place a one down the left side and a zero down the right, that is, construct $(P_{\mathbb{Z}_2}, 1, 0)$. Call this triangle $\widetilde{P}_{\mathbb{Z}_2}$. Note that when we identify all the odd permutations in P_{S_3} with one color and all the even permutations with another, we obtain $\widetilde{P}_{\mathbb{Z}_2}$ (the right picture in Figure 5). The reader can check that $\widetilde{P}_{\mathbb{Z}_2}$ is just $P_{\mathbb{Z}_2}$ with an extra edge of zeros down the right side.

Motivated by the success of the S_3 triangle, it seems reasonable to try other dihedral groups D_n. (Recall that $S_3 \cong D_3$.) By D_n, we mean the symmetry group of order $2n$ of the regular n-sided polygon. It is generated by a rotation ρ and a reflection σ subject to the relations $\rho^n = \sigma^2 = \iota$ and $\sigma\rho^k = \rho^{-k}\sigma$.

Exercise 3.1 *Construct the PascGalois triangles for the dihedral groups D_n where $n = 3, 4, 5, 6$, and 8. In each case, place the reflection σ down the left side of the triangle and the rotation ρ down the right. Are there any important properties that characterize these pictures? Focus on the patterns within the downward pointing subtriangles. Are they well-organized or chaotic? Can you think of any group properties that may be contributing to your observations?*

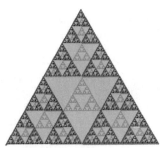 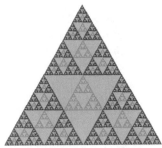

Figure 6: left: P_{D_4} with 129 rows;   right: P_{D_4} with 256 rows

The illustrations in Figure 6 show (P_{D_4}, σ, ρ). Note that the D_4 PascGalois triangle is much "sharper" (well-organized) than the $S_3 \cong D_3$ triangle. At first one may conjecture that the parity of n is the reason for this (e.g., P_{D_5} also has the disorganized downward pointing sub-triangles). In contrast, consider the D_6 and D_8 triangles in Figure 7. These two examples show that the parity of n is not the deciding factor. What appears to be important is whether or not D_n is a 2-group, that is, $n = 2^q$ for some positive integer q. (Recall that D_n is a p-group iff $p = 2$ and $n = 2^r$.) If D_n is a 2-group, then by Lagrange's Theorem all the group elements have order a power of 2. This produces the "sharp" sub-triangles. Dihedral groups that are not 2-groups contain elements whose orders are relatively prime to each other. These groups produce the more disorganized PascGalois triangles.

Figure 7: left: P_{D_6} with 129 rows;   right: P_{D_8} with 129 rows

4   Self-Similarity of $\mathbb{Z}_2 \times \mathbb{Z}_2$

In this section we characterize the self-similar growth structure of $P_{\mathbb{Z}_2 \times \mathbb{Z}_2}$. As we mentioned earlier, the self-similarity here is different from that of fractal geometry. Instead of zooming in to see copies of a pattern, we zoom out. In [10] and [6] the growth structure is studied for cyclic groups whose order is prime and prime power, respectively.

To study $P_{\mathbb{Z}_2 \times \mathbb{Z}_2}$, we first introduce some notation. Let T_m denote the first m rows (i.e., row 0 through row m-1) for $m \geq 1$ of $P_{\mathbb{Z}_2 \times \mathbb{Z}_2}$. As before $(0,1)$ is placed down the left side of the triangle and $(1,0)$ down the right. We place $(0,1)$ at the top of the triangle. We define T_m^* to be the same as T_m except that we put $(1,0)$ at the top. Finally, we define V_m to be the triangle in table 4.

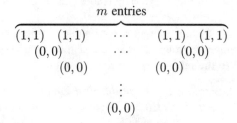

Table 4: Triangle V_m.

Exercise 4.1 *Construct $P_{\mathbb{Z}_2 \times \mathbb{Z}_2}$ with $(0,1)$ down the left side of the triangle and $(1,0)$ down the right. Are any patterns emerging? Are any pieces of the triangle repeated? Can you explain any patterns you find? How does the group structure play a role in this triangle?*

The following result solves Exercise 4.1. The triangle illustrated in this theorem is called a *growth triangle*.

Theorem 2 *For $n \geq 3$, the first 2^n rows of $P_{\mathbb{Z}_2 \times \mathbb{Z}_2}$ make a triangle that has the form found in Figure 8.*

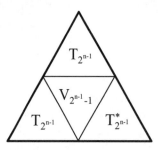

Figure 8: A growth triangle

Proof.   We proceed by induction on n. To simplify notation we write $a = (0,1)$, $b = (1,0)$, $c = (1,1)$, and $e = (0,0)$. For $n = 3$ we compute T_4 and T_8 directly (see Figure 9). Now assume the result is true for $n-1$, where $n \geq 4$. By the induction hypothesis we see that the last row of $T_{2^{n-1}}$ (i.e., row $2^{n-1} - 1$) is

$$\underbrace{a \quad b \quad a \quad b \quad \cdots \quad a \quad b \quad a \quad b}_{2^{n-1} \text{ entries}}$$

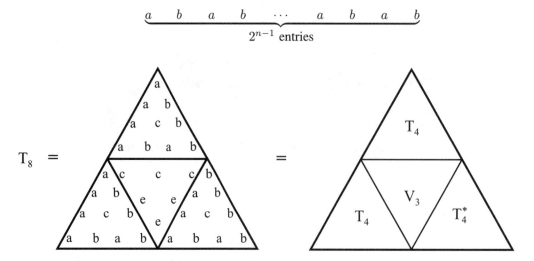

Figure 9: Triangles T_8 and T_4 (on top of T_8)

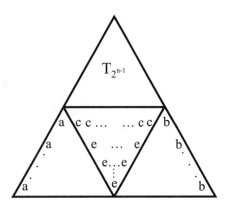

Figure 10: Portion of T_{2^n}

So row 2^{n-1} must be

$$a \qquad \underbrace{c \qquad c \qquad \cdots \qquad c \qquad c}_{2^{n-1} - 1 \text{ entries}} \qquad b$$

Since $c^2 = e$, part of T_{2^n} must be as indicated in Figure 10. Note that the middle triangle is just $V_{2^{n-1}-1}$. Then the rows 2^{n-1} to $2^n - 1$ satisfy what is indicated in Figure 11. Note that the left triangle has 2^{n-1} rows with a down the left side, b down the right, and a on top. Hence the triangle is $T_{2^{n-1}}$. The right triangle is the same except the top entry is b. So this triangle must be $T^*_{2^{n-1}}$. ∎

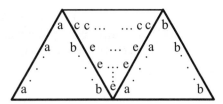

Figure 11: Rows 2^{n-1} to $2^n - 1$

Exercise 4.2 *Determine the growth triangles (see Theorem 2) for $P_{\mathbb{Z}_n}$ for $n = 2, 3, 4, 5,$ and 6. Compare the observed patterns with the subgroup lattice of each group. Then compare the orders of the group elements for each group versus the number of rows needed to repeat the self-similar structure. Does the value of n, whether it is prime or composite, seem to affect the corresponding PascGalois triangle?*

5   Conclusion

The pictures in this paper were drawn by the program *PascalGT* that we developed at Salisbury University using *Visual Basic*. A *Mathematica* implementation of our work has been written by Al Hibbard (Central College) and Ken Levasseur (UMass–Lowell). It can be found at [8]. Cynthia Woodburn (Pittsburg State University) has implemented some images on *Maple*. Her work can be found at [13]. Our package also allows the user to experiment with $\mathbb{Z}_n$ ring multiplication and any other binary operation on a finite set. The $\mathbb{Z}_n$ ring multiplication triangles are drawn by placing two elements a and b of $\mathbb{Z}_n \setminus \{0, 1\}$ down the sides of the triangle. Although the pictures are not as interesting, they can illustrate an important point to beginning abstract algebra students. The pictures for prime n look very similar to the group case. However, if n is composite then there exist $a, b \in \mathbb{Z}_n \setminus \{0, 1\}$ that cause zeros everywhere except for the sides of $P_{\mathbb{Z}_n}$. The reason is simple. If n is composite, then $\mathbb{Z}_n$ has zero divisors. Once there is a zero in the triangle, it generates an infinite sub-triangle of zeros beneath it. On the other hand, if n is prime then the nonzero elements of $\mathbb{Z}_n$ form a multiplicative group. Hence, zeros never appear in these triangles and we just obtain a group-generated triangle with 1 as the identity. The corresponding pictures illustrate the well-known theorem that $\mathbb{Z}_n$ is a field if and only if n is prime.

Some of the best student exercises from this project pertain to quotient groups. Quotient groups are arguably one of the most difficult topics in a first semester abstract algebra course. The notion of building a group out of equivalence classes can be unsettling. To make quotient groups more concrete we have students color all of the elements of a given coset the same color. This makes it easy for the student to see how all the elements in the coset are being identified to a single point in the quotient group. The following exercise is one such example.

Exercise 5.1 *View or construct the $P_{\mathbb{Z}_{10}}$ triangle. Note that there appear to be "competing patterns" overlapping one another. To understand why, consider the quotients $\mathbb{Z}_{10}/\{0, 5\}$ and $\mathbb{Z}_{10}/\{0, 2, 4, 6, 8\}$. For both quotient groups, find all the cosets. Then, in both cases, redraw a triangle with the elements of each coset colored the same. Do the two new triangles look familiar?*

Of course, when the student does this exercise they produce the $\mathbb{Z}_5$ and $\mathbb{Z}_2$ triangles. The point of the exercise is to help the students see $\mathbb{Z}_5$ and $\mathbb{Z}_2$ as quotients of $\mathbb{Z}_{10}$. This is a valuable assignment for other groups as well. The dihedral groups are especially worth trying. In this case, the student has to be sure to use left (right) cosets if left (right) multiplication is used to generate the triangle. This is what we did in section 3 with the $D_3 \cong S_3$ triangle. By coloring all the rotations one color and reflections another color, the isomorphism $S_3/A_3 \cong \mathbb{Z}_2$ is easily seen. All the dihedral groups have a normal rotational subgroup of index 2. However, other dihedral groups can have more than one normal subgroup. Thus, there may be several nice quotient triangles hidden inside P_{D_n}. Instructors should have their students find some of them.

Although we have included only a handful of exercises in this paper, there is a plethora of possible activities that can be done using the PascGalois triangles. For example, we have characterized many interesting properties in the $\mathbb{Z}_p \times \mathbb{Z}_p$ (p prime) triangle. In addition, we have studied automorphisms within the triangles for numerous groups. For those interested in applications to other areas, we are investigating connections with fractal geometry and cellular automata. Readers of this article are encouraged to contact the authors for more information on any of these topics.

Finally, we would like to note that other authors have recently sought to make modern algebra more visual. A perfect example is the book by Cox, Little, and O'Shea that relates polynomial algebra to geometric objects called varieties (see [4]). We hope that in the near future there will be more books and papers that emphasize the visual side of abstract algebra.

References

[1] M. Bardzell and K. Shannon, PascGalois web site, `http://faculty.salisbury.edu/~kmshannon/pascal/`

[2] J. Brown, Presentation on Pascal's Triangle at 1995 MATHCONN, Cedar Crest College, Allentown, Pennsylvania.

[3] S. Brown and D. Hathaway, *Fibonacci Powers and a Fascinating Triangle*, The College Mathematics Journal, **28** (1997), No. 2, 124–128.

[4] D. Cox, J. Little, and D. O'Shea, *Ideals, Varieties, and Algorithms*, 2nd ed., Springer-Verlag New York, Inc., 1997.

[5] H. Eves, *An Introduction to the History of Mathematics*, 6th ed., Saunders College Publishing, 1992.

[6] A. Granville, *Zaphod Beeblebrox's Brain and the Fifty-ninth row of Pascal's Triangle*, The American Mathematical Monthly, **99** (1992), 318-331.

[7] A. Granville, *Correction to: Zaphod Beeblebrox's Brain and the Fifty-ninth row of Pascal's Triangle*, The American Mathematical Monthly, **104** (1997), 848-851.

[8] A. Hibbard and K. Levasseur, web site for *Exploring Abstract Algebra with Mathematica*, `http://www.central.edu/eaam/grouppascal/`.

[9] P. Hilton, D. Holten, and J. Pedersen, *Mathematical Reflections*, Springer-Verlag New York, Inc., 1997.

[10] C. Long, *Pascal's Triangle Modulo p*, Fibonacci Quarterly, **19** (1981), 458-463.

[11] PascGalois library of images, `http://faculty.salisbury.edu/~kmshanno/pascal/MAANotes/library/`.

[12] D. Struik, *A Concise History of Mathematics*, 3rd ed., Dover Publications, Inc. New York, 1967.

[13] C. Woodburn, web site for *Maple* implementation of PascGalois, `http://www.pittstate.edu/math/Cynthia/pascal.html`.

Michael J. Bardzell, Department of Mathematics and Computer Science, Salisbury University, Salisbury, MD 21801; `mjbardzell@salisbury.edu`; `http://faculty.salisbury.edu/~mjbardze/`.

Kathleen M. Shannon, Department of Mathematics and Computer Science, Salisbury University, Salisbury, MD 21801; `kmshannon@salisbury.edu`; `http://faculty.salisbury.edu/~kmshanno/`.

Developing a Student Project in Abstract Algebra: The *Lights Out* Problem

John Wilson

Abstract. This article describes the development, solution, and results of a group project assigned during the second course of the abstract algebra sequence at Centre College. The *Lights Out* puzzle consists of 25 lights arranged in a square. When a light is touched, that light and each of the adjacent lights is toggled on or off. The students in this project were asked to develop a mathematical model for the puzzle and to determine how to turn off all of the lights starting with any given set of lights turned on. Several tips are given for faculty members who want to incorporate student projects into their courses.

1 Introduction

Centre College is a selective, liberal arts college in Danville, Kentucky enrolling a little over 1000 students. We typically have somewhere between 8 and 14 mathematics majors in each class. All mathematics majors are required to take a first course in abstract algebra dealing with group theory. This course is offered every fall. Many choose to take the follow-up course focusing on rings and fields that is offered alternate years in the spring. The first course usually has about 15 students. Most of the students in this class are junior mathematics majors with an occasional sophomore or senior and a few mathematics minors. The second course almost always has a mixture of 8 to 10 junior or senior mathematics majors. Many times I have asked the students to work on an individual or group project near the end of the second course. This article describes the development, solution, and results of one of these group projects. I have also included some general tips I have found to be helpful in assigning any individual or group project.

2 Development

Creating and assigning a good student project requires a great deal of thought and careful planning. Many questions have to be answered about a project before I will include it as part of my course. Is the problem interesting? Is the mathematics involved appropriate for the course? Is it possible for the student to get partial results? Is it possible for the student to extend or generalize the problem to get further results? How much will the project count in the course? How long will the students have to work on the project and how will they present their results? All of these are important aspects to consider when deciding to include a project as part of a course. In this section I will describe how the *Lights Out* project developed and how I included it in my abstract algebra course.

Tip #1: Ideas for projects come from many sources. Keep your eyes, ears and mind open. The idea for this project evolved slowly. Early in 1998 I was brainstorming with one of the mathematics students at Centre College about activities for our local student chapter of the MAA. Somehow we began to think about an electronic toy called *Lights Out* (manufactured by Tiger Electronics). The toy consists of 25 push-button square switches arranged in a 5×5 grid. Imagine this as a house with 25 rooms. Each room may have its light on or off. When a square is touched, the light in that room, and the lights in all rooms adjacent to it, are toggled on or off. (Thus, if a light was

125

on, it is turned off; similarly, if it was off, it is turned on.) Given a set of rooms with the lights on, the player is asked to find a set of switches that should be flipped to turn off the lights in all of the rooms.

I jotted down a few simplified puzzles to work on at our next meeting. Later that spring we used some similar puzzles for student activities at our MAA Kentucky section meeting. The students seemed to enjoy the problems. It became clear to me that further analysis of the *Lights Out* toy would make a nice student project in some setting.

Tip #2: Work through the project carefully and completely before you decide to assign it to students. During the summer of 1998 I developed the natural linear algebra model for the game. I first looked at the smaller 3×3 case with only 9 switches. The analysis turned out to be straightforward. Every pattern of rooms had a unique set of switches that turned off those lights. Next, I jumped to the 5×5 case that involved solving a system of 25 equations with 25 unknowns. The results in this case were surprising to me. These results further convinced me that this would be a good student project. The mathematics involved linear algebra over a finite field so I thought it would fit in well in the rings and fields course I was going to teach the next spring. It could also serve to remind the students of several linear algebra topics they had studied in their previous courses.

Tip #3: Write a clear description of the project. Before the start of the spring term, I typed a one-page description of the *Lights Out* project. This description clearly stated the goals and expectations for the students. I tried to give the students some guidance without making it simply a checklist of things to do. As with other favorite projects, this one had the typical three parts. The first section required the students to generate some examples or work on a simpler version of the problem. The second section required the students to solve the given problem. The third section asked the students to generalize or be creative in some way. The description of the *Lights Out* project I gave to my students is included here as an example.

Lights Out Project Description

In 1997, Tiger Electronics first marketed an electronic game called *Lights Out*. The game consists of 25 push-button square switches arranged in a 5×5 grid. Imagine this as a house with 25 rooms. Each room may have its light on or off. When a square is touched, the light in that room and the lights in all rooms adjacent to it are toggled on or off. The corner squares operate lights in three rooms each; the side squares operate lights in four rooms each; and the inside squares operate five rooms each.

The Problem. Given a set of rooms with the lights on, the player is asked to find a set of switches that should be flipped to turn off the lights in all of the rooms.

Goal. The goal of this project is to develop a mathematical model that can be used to analyze and solve the *Lights Out* problem.

Part I. Try to create a model for a smaller puzzle with 9 switches arranged in a 3×3 square instead of the 25 switches found in the *Lights Out* puzzle.

Use binary vectors to represent sets of switches or sets of lighted rooms. Investigate how the binary vector associated with a given set of rooms is affected when the switches in some given set are flipped. Further hints are available if needed. The model you end up with should allow you to answer the following questions for the smaller version of the *Lights Out* puzzle.

1. Given any set of lighted rooms, is it possible to find a set of switches that will turn off all of the lights?

2. Is it ever necessary to flip a particular switch more than once?

3. Does the order in which the switches are flipped make any difference?

4. Is there ever more than one set of switches that will turn all the lights off?

Part II. Use the ideas developed in Part I to describe some mathematical model for the 5×5 *Lights Out* puzzle. Are there any substantial differences in the two models and the resulting solutions?

Part III. Design your own puzzle similar to the *Lights Out* puzzle and create a mathematical model to analyze it.

3 Solution

In this section I will present the mathematical model I used during the summer of 1998 to solve the *Lights Out* problem.

Tip #4: Teach problem solving at every opportunity. One common problem solving strategy in mathematics is to look first at a smaller version of the problem that is to be solved. In this case we will consider a game similar to the *Lights Out* game but having only nine switches arranged in a 3×3 grid as shown in Figure 1. To develop a mathematical model of the game we can represent every individual switch by a binary vector of length nine. Switch 1 is represented by $[1, 0, 0, 0, 0, 0, 0, 0, 0]$. The ith switch is represented by the binary vector with a single 1 in the ith location and zeros elsewhere. These nine vectors are called the basic switch vectors. Any set of switches may be represented by a binary vector called a switch vector of length 9. For example the switch vector $[1, 1, 1, 0, 0, 0, 0, 0, 0]$ corresponds to the set of switches $\{1, 2, 3\}$.

Figure 1: The 3×3 game grid.

Associated with each set of switches is a set of rooms controlled by those switches. In the mathematical model we can associate with each switch vector another binary vector of length 9 called a room vector. A room vector has a 1 in each location where the light is toggled. For example switch 1 controls the set of rooms $\{1, 2, 4\}$ so the basic switch vector $[1, 0, 0, 0, 0, 0, 0, 0, 0]$ is associated with the room vector $[1, 1, 0, 1, 0, 0, 0, 0, 0]$. The room vectors associated with the nine basic switch vectors are called the basic room vectors.

A 9×9 binary matrix called the On matrix is created using the nine basic room vectors as the columns. This matrix is shown in the example below. The On matrix provides the means to find which rooms are affected by pressing any given set of switches. Mathematically, we can calculate an unknown room vector, r, associated with a given switch vector, s, by multiplying the transpose of s by the On matrix.

Example. Our goal is to find the set of rooms toggled by flipping the switches in the set $\{1, 2, 3\}$. The switch vector is given by $s = [1, 1, 1, 0, 0, 0, 0, 0, 0]$. The corresponding room vector is calculated by the matrix product $\text{On} * s^T = r^T$ (where s^T is the transpose of s) with the arithmetic being done over the finite field $GF(2)$ (see Equation 1).

$$\begin{bmatrix} 1 & 1 & 0 & 1 & 0 & 0 & 0 & 0 & 0 \\ 1 & 1 & 1 & 0 & 1 & 0 & 0 & 0 & 0 \\ 0 & 1 & 1 & 0 & 0 & 1 & 0 & 0 & 0 \\ 1 & 0 & 0 & 1 & 1 & 0 & 1 & 0 & 0 \\ 0 & 1 & 0 & 1 & 1 & 1 & 0 & 1 & 0 \\ 0 & 0 & 1 & 0 & 1 & 1 & 0 & 0 & 1 \\ 0 & 0 & 0 & 1 & 0 & 0 & 1 & 1 & 0 \\ 0 & 0 & 0 & 0 & 1 & 0 & 1 & 1 & 1 \\ 0 & 0 & 0 & 0 & 0 & 1 & 0 & 1 & 1 \end{bmatrix} * \begin{bmatrix} 1 \\ 1 \\ 1 \\ 0 \\ 0 \\ 0 \\ 0 \\ 0 \\ 0 \end{bmatrix} = \begin{bmatrix} 0 \\ 1 \\ 0 \\ 1 \\ 1 \\ 1 \\ 0 \\ 0 \\ 0 \end{bmatrix} \tag{1}$$

Since the room vector is $[0, 1, 0, 1, 1, 1, 0, 0, 0]$ we conclude that the set of rooms $\{2, 4, 5, 6\}$ are toggled by the set of switches $\{1, 2, 3\}$. This example illustrates how it is possible to determine the set of rooms toggled by a given set of switches but the problem that we must solve in the game is to determine a set of switches that will toggle a given set of rooms. Mathematically, we must solve the matrix equation $\text{On} * s^T = r^T$ where r is a known room vector and s is an unknown switch vector. If the On matrix is invertible over the field $GF(2)$ then this equation is easy to solve. Using the command $\text{Off:=Inverse(On)} \bmod 2$ with *Maple*'s linear algebra package allows us to quickly determine that this On matrix is in fact invertible over $GF(2)$. I called this inverse the Off matrix.

Example. Suppose the lights are on in the rooms $\{1, 2, 5, 7\}$. The room vector is given by $r = [1, 1, 0, 0, 1, 0, 1, 0, 0]$. The corresponding switch vector is computed using the matrix equation $s^T = \text{Off} * r^T$ (see Equation 2). Since the switch vector is $[0, 0, 1, 1, 0, 1, 1, 1, 0]$ we conclude that the lights in rooms $\{1, 2, 5, 7\}$ are turned off using the switches in $\{3, 4, 6, 7, 8\}$.

$$
\begin{bmatrix} 0 \\ 0 \\ 1 \\ 1 \\ 0 \\ 1 \\ 1 \\ 1 \\ 0 \end{bmatrix}
=
\begin{bmatrix}
1 & 0 & 1 & 0 & 0 & 1 & 1 & 1 & 0 \\
0 & 0 & 0 & 0 & 1 & 0 & 1 & 1 & 1 \\
1 & 0 & 1 & 1 & 0 & 0 & 0 & 1 & 1 \\
0 & 0 & 1 & 0 & 1 & 1 & 0 & 0 & 1 \\
0 & 1 & 0 & 1 & 1 & 1 & 0 & 1 & 0 \\
1 & 0 & 0 & 1 & 1 & 0 & 1 & 0 & 0 \\
1 & 1 & 0 & 0 & 0 & 1 & 1 & 0 & 1 \\
1 & 1 & 1 & 0 & 1 & 0 & 0 & 0 & 0 \\
0 & 1 & 1 & 1 & 0 & 0 & 1 & 0 & 1
\end{bmatrix}
*
\begin{bmatrix} 1 \\ 1 \\ 0 \\ 0 \\ 1 \\ 0 \\ 1 \\ 0 \\ 0 \end{bmatrix}
\tag{2}
$$

The 3×3 puzzle is now completely solved. The fact that the On matrix has an inverse allows us find a unique set of switches to turn off the lights in any set of rooms. The order in which the switches are pressed makes no difference. Since the switches simply toggle the lights on and off, pressing a switch an even number of times has the same effect as doing nothing. Flipping a switch an odd number of times is the same as flipping it once.

The students working on the project should now be confident in how to proceed to Part II of the project. The analysis of the 5×5 *Lights Out* game is carried out exactly the same as the 3×3 case. A 25×25 binary matrix I have called the LightsOn matrix is created with columns being the basic room vectors. Again we need to solve the matrix equation $\text{LightsOn} * s^T = r^T$ for a switch vector, s, given any room vector, r. After entering the LightsOn matrix in *Maple*, I again used the Inverse command to try to compute the inverse over $GF(2)$. I was very surprised to find that this matrix was not invertible. By using the command $\text{Gaussjord(LightsOn)} \bmod 2$ it is clear that the rank is only 23.

This unexpected twist in the project forces the students to conclude that some sets of rooms can never be turned off. This in turn opens the door to many new questions. Is it possible to characterize which sets of rooms have a solution? What is the probability that a randomly selected set of rooms will have a solution? Is there some other interesting way of wiring the switches so that the 5×5 game will have a unique solution for every set of rooms? Are there sizes other than the 3×3 game for which the LightsOn matrix is invertible? This type of questioning is one of the main reasons for doing projects in the first place. At this point students begin to see the real excitement in doing mathematics.

The goal of Part III of the project is to tap into this enthusiasm. I asked the students to design a new but similar puzzle and carry out the mathematical analysis. I chose to go back to the modeling aspect of the project. Another direction to follow would be to investigate other matrices similar to the LightsOn matrices.

4 Results

In the spring of 1999 I divided my abstract algebra class into four teams consisting of two or three students. I gave a description of the project to one of the teams at midterm. I required no written report. Instead, they were to present their results at our last class meeting of the year. They were completely responsible for that entire class meeting. I directed them to a web site that had a *Java* applet version of the game and allowed them to borrow my game. They also had access to the *Maple* software on our campus computer network.

Tip #5: Assign the project early in the term and insist on several progress reports. Throughout the rest of the term the project progressed about as I had expected. The students came by my office three or four times to discuss their progress and ask appropriate questions. Some questions were theoretical in nature dealing with linear algebra topics. Others were technical issues related to using *Maple*. (Commands such as Inverse and Gaussjord must start with upper case letters if one wants to do the arithmetic modulo 2.)

The new puzzle they chose to analyze in Part III of the project was a 3×3 version with different intensities or colors of light. The switches were wired as before but instead of just toggling the lights on and off the lights cycled through a prescribed set of intensities (off, low and high). This generalization was easy to handle with *Maple* by simply doing the arithmetic mod 3. They wrote a computer program allowing the students to select how many intensity levels were available. These intensity levels were represented by different colors for the rooms. All of the students in the class seemed to enjoy the presentation.

I consider this project to be one of the best I have had students work on in my abstract algebra course. My hope is that some of the students who see this problem will want to continue the investigation in directions of their own choice.

References

Several articles have been written about using mathematics to analyze puzzles similar to the *Lights Out* puzzle.

[1] M. Anderson and T. Feil, *Turning lights out with linear algebra*, Mathematics Magazine **71** (1998), 300–303.

[2] B. Lotto, *It all adds up to elegance and power*, Vassar Quarterly, Winter 1996, 18–23.

[3] J. Missigman and R. Weida, *An easy solution to Mini Lights Out*, Mathematics Magazine **74** (2001), 57–59.

[4] D. Pelletier, *Merlin's Magic Square*, Amer. Math. Monthly **94** (1987), 143–150.

[5] D. Stock, *Merlin's Magic Square revisited*, Amer. Math. Monthly **96** (1989), 608–610.

John Wilson, Department of Mathematics, Centre College, Danville, KY 40422; wilson@centre.edu.

Learning Permutation Group Theory via Puzzles

John O. Kiltinen

Abstract. This article describes the author's use of computerized puzzles for helping students gain a deeper understanding of the theory of permutation groups in his abstract algebra course. The software, which he has developed using a cross-platform (Macintosh and Windows) development tool called *MetaCard*, presents several puzzles to the student that require aspects of group theory (such as commutators and conjugates) to obtain a solution. The puzzles range from a simple one that creates an environment for exploring transpositions to several very challenging ones that are modeled after commercial physical puzzles. All of the puzzles are two-dimensional for ease in understanding and to avoid orientation issues such as those that arise with *Rubik's Cube*. The article describes the puzzles and their computerized implementation, outlines the mathematical ideas that they illuminate, and describes the use the author has made of them in teaching his introductory undergraduate abstract algebra course.

1 Introduction

Many will remember the popularity in the late 1970s and early 1980s of *Rubik's Cube* and the variety of similar puzzles that it inspired. These puzzles, which involve rearranging pieces subject to the particular physical constraints imposed by the puzzle, offer concrete representations of subgroups of permutation groups. For this reason, this author, as well as many other teachers of undergraduate abstract algebra courses, have incorporated discussions of them into their courses.

One of the puzzles that followed *Rubik's Cube* is called *Top Spin*. It has twenty numbered disks that can revolve freely on an oval-shaped track. There is a turntable at the top of the track that holds four disks at a time. By rotating this turntable through 180 degrees, one can reverse the order of the four disks that are on it. The object of the game is to scramble the disks and then restore them to the original order by a series of slides of disks around the track and rotations of the turntable.

Top Spin is a pure permutation group puzzle, in contrast to *Rubik's Cube*, which raises issues of the orientation of its pieces in addition to their placement. This makes it a simpler tool for exploring ideas of the theory of permutation groups. The fact that it is two-dimensional rather than three also makes it easier to understand.

The arrival of *Top Spin* on the market coincided with the appearance of the *Hypercard* software development system for the Macintosh computer. *Hypercard* provided the desirable combination of programming tools and a graphical user interface in an easy-to-use development environment. These circumstances induced the author to develop a *Hypercard* stack that simulated and generalized the *Top Spin* puzzle. The development of this computerized puzzle inspired ideas for several other puzzles, which the author has also implemented. He has used these puzzles in the teaching of his undergraduate abstract algebra course for about ten years. Currently, the puzzles have been upgraded using a cross-platform development system called *MetaCard*.

The purpose of this article is to describe the puzzles, briefly discuss their computer implementation, outline the mathematical ideas they illuminate, and review the use the author has made of them in teaching his introductory undergraduate abstract algebra course.

2   Description of the Puzzles

2.1   *Oval Track* puzzle

Figure 1 shows the user-interface screen for the *Oval Track* puzzle, the one based upon *Top Spin*. One can move the numbered disks around the track either by dragging disks on the track or by pressing the numbered buttons in the lower-left corner. One turns the turntable by pressing the Do Arrows button. The Scramble button with the die icon presents the user with several options regarding how to scramble the disks. It will scramble them and then the user can work to get every numbered disk back to its starting place.

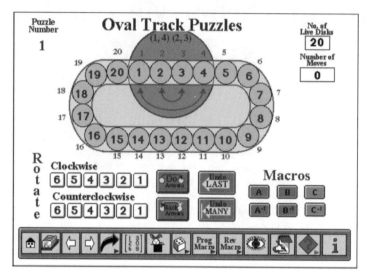

Figure 1: The Oval Track Puzzle.

The arrow keys on the left of the button bar at the bottom of the window are for navigating to other versions of the puzzle that replace the small turntable permutation $(1, 4)(2, 3)$ with a different small permutation. For example, some other permutations are $(3, 2, 1)$, $(4, 3, 2, 1)$, $(5, 4, 3, 2, 1)$, $(1, 2)(3, 4)$, and $(1, 3)(2, 4)$. There are 19 variations in all. In each case, the permutation is shown with arrows, and the Do Arrows button performs it. The Back Arrows button performs the inverse of this permutation, literally following the arrows in a backwards direction.

The more advanced functions that the puzzle offers include the capability of recording and repeating a series of commands by the press of a single button. Pressing the Prog Macro button initiates the process of entering a program into any of the three buttons labeled "A", "B", or "C", which the user completes by simply performing the steps to be recorded. Once the sequence of steps is recorded, the user can repeat it by pressing the button to which the macro was assigned. The inverse of the programmed permutation is obtained by pressing the corresponding inverse button (such as A^{-1}).

Another variation that the puzzle allows is setting the "live number" (number of active disks) to anything from the full 20 disks down to the smallest number n such that the Do Arrows permutation fits into the interval $[1, n]$. When the active number n is less than 20, the locations with numbers in the interval $[n + 1, 20]$ have X's in them, and a rotation will have the numbered disks jump across these deactivated locations.

The program records the entire sequence of moves that the user has entered since the last Start Over, and allows the user to reverse a sequence of steps. The Undo Last button reverses the last step entered, and the Undo Many button brings up a dialog box that allows the user to pick how many of the most recent moves to undo.

2.2   *Transpose* puzzle

The *Oval Track* puzzles in all their variations are presented in one window. A second package, called *Multi-Puzzle*, presents another set of three puzzles in a separate window. The first of these other three is a puzzle called the *Transpose* puzzle; see Figure 2. It has the simple purpose of allowing the user to gain concrete experience with the fact that any permutation of the set of integers from 1 to n can be obtained by a sequence of transpositions. It presents the user with a grid of numbered boxes, each containing a numbered disk. The only way to move the disks

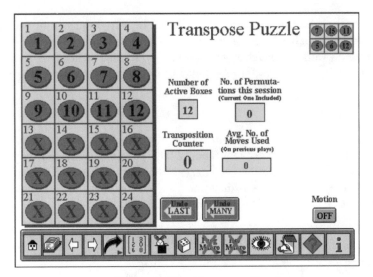

Figure 2: The Transpose Puzzle.

around is by means of transpositions, or swapping two disks between two boxes. One does this by dragging a disk from its present location to the box where one wants it. After dragging a box, upon releasing the mouse button, the disks in the source and target boxes exchange places. The functions of the other buttons for this puzzle correspond naturally with their names (see Figure 2).

It does not take much experience with repeated swapping of disks between boxes before the user truly knows that any permutation of the disks can be produced by this process. Conversely, if the user has the computer scramble the disks, the user can easily learn to get them all back to their starting locations by means of transpositions, and can discover several different algorithms for doing this. The user who knows a bit of the theory of permutation groups may also discover the relationship between the number of transpositions used for `solving` a permutation and its cycle structure.

2.3 *Slide* puzzle

The second puzzle presented by *Multi-Puzzle* is called the *Slide* puzzle. This one is based on the very old and familiar "15" puzzle that has 15 numbered squares on a 4×4 square with one empty spot. In this computer realization, one moves a numbered disk from an adjacent square into the blank spot by clicking on that numbered disk, whereupon it exchanges places with the blank spot.

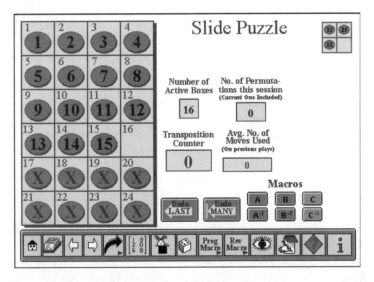

Figure 3: The Slide Puzzle.

The puzzle allows for many variations by altering the number of active squares, by building obstacles out of deactivated squares, and by having two blank spots rather than one. It also offers the options of vertical or horizontal wrapping, allowing the puzzle to work as though it were on the surface of a cylinder or a torus. The puzzle has a scramble option, and, like *Oval Tracks*, allows for programming macro buttons and undoing steps.

2.4 *Hungarian Rings* puzzle

The final puzzle in this set of three is called the *Hungarian Rings* puzzle. This too is modeled after a plastic puzzle that was available in the 1980s. This computer realization of the puzzle is similar to the *Oval Tracks* puzzle in nearly all respects. One difference, however, is that the puzzle can be put into a four-color mode, as shown in Figure 4, or into a number mode where each disk has a distinct number. The four-color mode is modeled after the physical puzzle, and the number mode extends the challenge and aids in analysis.

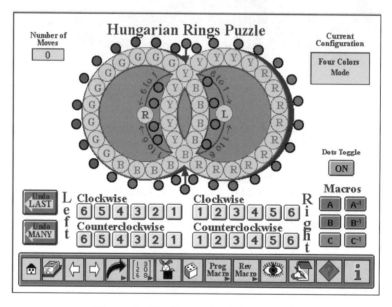

Figure 4: The Hungarian Rings Puzzle.

2.5 Common Features

All of the puzzles make use of sound to confirm for the user that mouse movements and button clicks have been interpreted by the computer as intended. The puzzles are driven by a standalone application program called *Puzzle Driver* that is available for either the Macintosh or the Microsoft Windows operating systems. The package also includes a `Puzzle Resources` component that offers the user a collection of preset configurations and macro programs, and provides a place for creating and collecting additional resources for future use. From this window, the user can also download the latest information about the puzzles over the internet.

The components of this package are built using the *MetaCard* system, which is a powerful cross-platform software development system. *MetaCard* provides the capability of implementation on the Unix platform also, but the author has not yet done this.

3 The Mathematical Ideas

These puzzles provide a context for developing a concrete understanding of certain aspects of the theory of permutations groups. The simplest of them, the *Transpose* puzzle, gives the student hands-on experience with the principle that every permutation can be expressed as a sequence of transpositions. The puzzle also provides a natural context for discovering and comparing algorithms for expressing a permutation in terms of transpositions, for discovering ways of telling how many transpositions are needed for a given permutation, and for asking about the average number of transpositions needed. The *Transpose* puzzle also lays the groundwork for an understanding of the Parity Theorem.

The "end-game" arrives when one has just a few pieces remaining to be put into their proper places. Typically one finds that as one tries to get the last pieces into place, the common sense strategies that were adequate earlier are now insufficient, causing unintended displacements along with each of the pieces put right. The theory of permutation groups provides tools for developing effective end-game strategies, as we will illustrate below.

The Parity Theorem comes into play frequently to frame what is possible and what is impossible. A simple example comes in the first *Oval Track* puzzle shown in Figure 1. If an odd number of disks (n) is active, then the generating permutations are the n-cycle $(1, 2, 3, \ldots, n)$ and the turntable permutation $(1, 4)(2, 3)$. Since both of these permutations are even, all combinations of them are also even. This means that if the random scrambler produces an odd permutation, the user will not be able to solve it completely. The perceptive student has ample opportunity to see many applications of parity ideas in analyzing the puzzles.

The puzzles also provide useful insights into the role of commutators and conjugates. A *commutator* is a group element of the form $ABA^{-1}B^{-1}$. If the elements A and B commute, the commutator $ABA^{-1}B^{-1}$ is just the identity element of the group. This being the case, in a permutation group, if one of the elements, say B, moves only a small set of elements, then a commutator involving B should differ from the identity permutation by a small amount. That is, a commutator ought to move just a few elements.

In exploring these puzzles, one wants to look for commutators that produce small, controlled changes that one can use to put the final few pieces into their right places. Discovering these commutator tools is integral to having an efficient strategy for solving a puzzle.

We extend the reach of our fine-control commutators by means of conjugates. For two elements A and B in a group, we call the element ABA^{-1} the *conjugate* of B by A. An important fact is that conjugates in permutation groups share the same cycle structure. Thus, for example, if B is a small permutation (e.g., a three-cycle) that is useful for an end-game, then no matter how complicated the permutation A is, the conjugate ABA^{-1} of B by A is also a three-cycle. Thus, if one has discovered a useful three-cycle, one can use it to get other three-cycles.

To illustrate, with the first *Oval Track* puzzle, using R to denote a one-position clockwise rotation of the disks and T to denote a half-revolution of the turntable, one can check that

$$R^{-3}TR^3T^{-1} = (1, 4, 7)(2, 3)(5, 6).$$

(Recall that we are performing the operations in a left-to-right manner.) That is, this particular commutator produces a permutation that decomposes into a three-cycle and two transpositions. If one repeats the motions, the transpositions return their elements to where they came from, and the three-cycle is squared, which is again a three-cycle. That is, $B = (R^{-3}TR^3T^{-1})^2 = (1,7,4)$. We have a basic three-cycle, which we could program into button B for ease of future use.

Now suppose one wants to consider some other three-cycle (r, s, t), to move the disk numbered r to disk s, disk s to disk t, and disk t to r. We can get this three-cycle as a conjugate of B as follows. First, create a "set-up procedure" (call it A) that puts the disks r, s, and t into spots 1, 7, and 4, respectively. Since we will later have to reverse this procedure by performing the inverses of each step in the opposite order, we can program the procedure into button A to avoid having to remember or write down the steps. Once A is entered, then perform procedure B. Finally, follow this by the procedure A^{-1} (using the A^{-1} button). We then have $(r, s, t) = ABA^{-1}$.

A little experience with the puzzle makes it clear that it affords enough freedom to get any three disks into spots 1, 7, and 4. Thus, one can get any three-cycle whatever as a conjugate of the basic one we have discovered. This insight, together with another result from the theory of permutation groups, goes a long way toward showing how solvable the puzzle is. A theorem concerning permutations says that every even permutation can be expressed as a sequence of three-cycles. Thus, if we can perform all three-cycles, then we can perform all even permutations.

With a similar line of reasoning, one can observe that if one can identify a way of getting a single transposition on the puzzle, then one can get all other transpositions as conjugates of the basic one. This insight, together with what one learned from the *Transpose* puzzle, namely that all permutations can be expressed with transpositions, we know that we can totally solve the puzzle once we have found a way to perform a single transposition.

As illustrated by these examples, the puzzles make relevant and concrete some of the fundamental ideas of the theory of permutation groups. One gets a hands-on feeling for some practical consequences of the results.

4   Pedagogical Matters

As mentioned in the introduction, the author has used these puzzles for about ten years in teaching the introductory undergraduate abstract algebra course at Northern Michigan University. The class spends several sessions together in a computer lab working on the puzzles while having the instructor there to answer questions. Beyond that, they are expected to work on the puzzles on their own time in order to develop their understanding. Two extra class sessions are spent discussing the puzzles in a regular classroom (with a computer and projector available to demonstrate).

In addition to using these puzzles, the students make use of the group theory package of the *Maple* computer algebra system. *Maple*'s group theory package includes routines that allow it to calculate the order of subgroups of the group of permutations on a finite set $\{1, 2, \ldots, n\}$. One defines the subgroup for *Maple* by specifying its generators, then *Maple* can determine the number of elements in the subgroup. This information allows the student to become aware that changes in the number of active objects can result in either subtle or unexpectedly dramatic changes in the degree of solvability of the puzzle. For some, having this information motivates a desire to understand why this is so.

The students have about two weeks to work on the puzzles on their own time. They are then asked to submit the first of two written reports of their results. In the first preliminary report, they answer questions that probe their understanding of the *Transpose* puzzle and the issues it raises. They are also asked to report on their early findings with the *Oval Track* puzzle. This report is read and returned with encouragement for continued diligence. The instructor points out when they are headed in a productive direction or when they need to shift directions.

The second report is due at the end of the semester. This allows the students several more weeks to mull over the ideas and to work with the puzzles. By the time of the second report, the students are expected to have developed an appreciation for the power and generality of the conjugacy idea, and to recognize parity issues when they arise. After reading these second reports, the instructor holds one-on-one sessions with the students at a computer to allow them to demonstrate their understanding. By showing students what they can do with the puzzles, they are often able to demonstrate more insight than they were able to indicate in writing. The dialog also provides an opportunity for sharing ideas which help them to consolidate their understanding.

Some students seem to genuinely enjoy the work with the puzzles. Others receive it less enthusiastically, perhaps because the activity requires spending more time-on-task in order to gain the expected level of understanding than they are interested in spending.

The level of student interest has grown as a result of the additions of color and many new features that the author made while converting the software from *Hypercard* to *MetaCard*. The newer versions run faster, offer more automation, and the color makes it more pleasant to work with the puzzles longer.

The *Hungarian Rings* puzzle is the most difficult one, and students are told this from the start. However, the author has noted with surprise that students are gravitating to this puzzle more quickly now that color has been added.

5   Conclusion

One of the goals that the author has for this software is to give students sufficient concrete experience to provide a basis for them to engage in meaningful abstract work. Anecdotal evidence suggests that the experience indeed works this way for some students. Recently a student told about how after the role of conjugacy was explained to him in front of the computer in a lab session, he was able to work out all of the details of an end-game strategy with pencil and paper. The author regards this as evidence that he had really internalized the important ideas.

The author is in the process of writing a book that will provide instruction about the puzzles and the group theory that one needs to understand them. He hopes to have the book marketed along with the software. The interested reader can contact the author for information on the progress of this project.

John O. Kiltinen, Department of Mathematics and Computer Science, Northern Michigan University, Marquette, MI 49855; kiltinen@nmu.edu; http://www.nmu.edu/mathpuzzles/.

Ringing the Changes: An Aural Permutation Group

Lucy Dechéne

Abstract. Change ringing is the British sport of ringing all possible permutations of a set of n tuned bells following six rules. Historically, British bell ringers discovered permutation groups nearly a century before mathematicians. A group of permutations that follows the rules (called an extent) can be created by finding certain cosets of a subgroup of S_n. Creating extents is a fun way of practicing multiplication of permutations and exploring permutation groups, subgroups, the idea of algebraic words, and cosets. A theorem by A. T. White elegantly relates the existence of an extent to the existence of a Hamiltonian circuit in a Cayley graph. Some suggested exercises are given.

1 Introduction

I have taught an abstract algebra course for mathematics majors periodically over the last 23 years. Recently I have also taught a graduate course in abstract algebra for high school teachers in our M.A.T. program. I have always been on the lookout for concrete and fun examples of cosets. Since I loved the sound of bells, I became a carillonneur as a graduate student. It wasn't until I spent a sabbatical in England and interacted with change ringers that I realized it would be fairly easy to combine bells and cosets of permutation groups for a fun and novel application by non-mathematicians. Therefore, for the last several years I've taken an aural approach to reinforcing the ideas of factor groups of permutation groups.

The English sport of change ringing arose in the 17th century. A *change* is the ringing of all n tuned bells in a bell tower in one of $n!$ ways. A collection of such tuned bells in a tower is a *ring*. If the bells are numbered according to descending pitch, bell 1 (the highest) is the *treble* and bell n is the *tenor*. A *round* is the ringing of the bells in sequence from highest to lowest. Intricate methods called *extents* [13, 10] have been developed for ringing all possible changes on n bells. Many of these extents depend on group theory. Thus, change ringing is an excellent way to introduce permutation groups and cosets.

1.1 Rise of Change Ringing

Bells have played an important role in England since before William the Conqueror [1]. By the time Henry VIII closed Catholic churches and monasteries, most religious institutions had sets of bells used to signal the populace. Bands of ringers at that time had the job of ringing the whole set of bells for special occasions. The churchwarden's account book for St. Margaret's in Westminster noted [5]:

1586	Paid for ringing at the beheading of the Queen of Scots	1 shilling
1605	Paid the ringers for ringing at the time when the	
	Parliament Hous [sic] should have been blown up	10 shillings

After the Reformation, English bells started to be hung on a full wheel, which allowed a 360-degree rotation [1, 5]. The further refinement of a slider and a stay allowed a ringer to set the bell in the mouth up position. This permits the ringer to temporarily halt the bell and restart it precisely. Thus, ringing bells in a precise way for change ringing became possible in the early 17th century [5, 8]. In 1637, the Society of College Youths was formed to ring

London bells [5, 4]. At first the members only rang simple rounds and a few set changes. About the year 1642, *plain changes* involving only the transposition of two bells at a time were tried [5]. In 1657, more complicated extents such as *cross changes* involving double transpositions were tried [1]. By 1668, printer and change ringer Fabian Stedman had worked out various extents that were essentially based on properties of permutation groups and cosets nearly a century before Lagrange and other mathematicians discovered the same concepts [8, 3, 13].

1.2   Fabian Stedman

Fabian Stedman (1641–1713) was apprenticed to Master Printer Daniel Pakeman of London in 1655 [4]. In 1664 he joined the Society of College Youths [13]. In 1668 he had Rev. Richard Duckworth write the book *Tintinnalogia*, which set forth Stedman's explanations of change ringing and the rules for many types of extents [3]. Then, in a modern move, Stedman set out on horseback on a book tour to Kettering, Leicester, and the Midlands to promote the book and encourage ringers outside of London to embrace more complicated extents [5]. In 1677, he became the Steward of College Youths [13]. This was the same year he wrote a sequel to the original book, *Campanalogia* [8]. Both books show a good intuitive knowledge of how to generate permutation groups without giving any mathematical definitions.

2   The Abstract Algebra of Change Ringing

2.1   Stedman and Factorials

In both books Stedman shows an awareness of how many changes there are for n bells. In fact, he gets very carried away with giving examples to underscore how huge the numbers are. I like to share some of his passages with my students because they cast a light on both the historical period and the vocation of printer in the 17th century. A good example is the following passage [8], which puts into human terms 12! (the number of changes on twelve bells).

> And if twelve men should attempt to ring all these changes on twelve bells, they could not effect it in less than seventy-five years, twelve lunar months, one week, and three days, notwithstanding they ring without intermission, and after the proportion of 720 changes every hour. Or, if one man should attempt to prick [write] them down on paper, he could not effect it in less than the aforesaid space. And 1440 being pricked on a sheet, they would take up six hundred sixty-five reams of paper and upwards, reckoning five hundred sheets to ream; which paper at five shillings the ream, would cost one hundred sixty-six pounds five shillings.

2.2   Requirements for an Extent

The rules for an extent that were in Stedman's books and are still in use today are [8, 3, 13]:

1. The first and last changes are both rounds.

2. No other change is repeated; thus, every other change is rung exactly once.

3. From one change to the next, no bell changes its order of ringing by more than one position.

There are three additional rules that are required for modern extents [13, 10, 11].

4. In the plain course, no bell occupies the same position in its order of ringing for more than two successive changes.

5. The *working bells* do the same work in the plain course.

6. Each *lead* (division) of the extent is palindromic in the transitions employed.

2.3   Creating Plain Bob Minimus from D_4

In the terminology of Stedman, as well as contemporary terminology, a bell *hunts up* if it goes from first to last place through a series of transpositions. A bell *hunts down* if it works from last to first place by a series of transpositions [8, 3, 10]. The extent of Plain Bob Minimus is called "Plain Bob" after the type of method and "Minimus" because it occurs on 4 bells. In Plain Bob Minimus, the *working bells* are 2, 3, and 4. That is, all except the treble, which is hunting.

Change ringers describe the Plain Bob in a variety of ways. One description involves the fact that the treble bell hunts up, stays in place one change, and then hunts down. It stays in place one change, and then repeats the same pattern twice. In the mean time, some other bells are switching places in the order of ringing.

A clearer explanation by change ringers of Plain Bob on an even number of bells is [4]:

1. All pairs change.

2. All inner pairs change.

3. At the end of a lead, to start another lead, the bells in the 1st two positions do not change and all other pairs do.

From a perspective of permutations, on four bells, the pattern in a lead is to do $(12)(34)$ followed by (23) [10, 6]. The first lead in Plain Bob Minimus is given in Table 1. Note that if we apply the permutation (23) again, we get the identity permutation. Thus, we have completely generated a group isomorphic to D_4, the symmetries of a square, but we haven't completely generated S_4, which is our goal. A.T. White, the source of the mathematics in this paper, suggests we call the group derived in this fashion the *hunting group*, since the treble bell was hunting [13, 10, 11]. In this article, we use H to denote the hunting group.

Bell order	Permutation applied	Resulting permutation product
1 2 3 4	(1)	(1)
2 **1** 4 3	$(12)(34)$	$(12)(34)$
2 4 **1** 3	(23)	$(23)(12)(34) = (1342)$
4 2 3 **1**	$(12)(34)$	$(12)(34)(1342) = (14)$
4 3 2 **1**	(23)	$(23)(14)$
3 4 **1** 2	$(12)(34)$	$(12)(34)(23)(14) = (13)(24)$
3 **1** 4 2	(23)	$(23)(13)(24) = (1243)$
1 3 2 4	$(12)(34)$	$(12)(34)(1243) = (23)$

Table 1: First lead in Plain Bob.

In trying to solve this problem, Stedman and other change ringers stumbled upon what was essentially the idea of cosets. Rule (3) above (a transition rule) starts the process of creating the coset $H((12)(34)(23))^3(12)(34)(34)$ or $H(234)$. When we finish the next lead, we will have finished that coset. Then we follow the transition rule again and create the final coset $H(234)^2$, which will complete S_4. Thus, the second and third leads are as in Table 2.

1 3 4 2	**1** 4 2 3
3 **1** 2 4	4 **1** 3 2
3 2 **1** 4	4 3 **1** 2
2 3 4 **1**	3 4 2 **1**
2 4 3 **1**	3 2 4 **1**
4 2 **1** 3	2 3 **1** 4
4 **1** 2 3	2 **1** 3 4
1 4 3 2	**1** 2 4 3
	1 2 3 4

Table 2: Second and third leads.

2.4 Why the Extent Meets Conditions

White suggests in [13] that we call $a = (12)(34)$, $b = (23)$, and $c = (34)$. Then

$$H = \{e, a, ab, aba, abab, ababa, ababab, abababa\}.$$

If we let $w = abababac = (ab)^3ac$, then the next lead is $\{w, wa, wab, waba, wabab, wababa, wababab, wabababa\}$. Multiplying by w again to get the last lead, we have

$$\{ww, wwa, wwab, wwaba, wwabab, wwababa, wwababab, wwabababa\}.$$

(Note that in White's notation, ab means a followed by b. So it is actually permutation multiplication on the left by b.)

He notes in [13] that the extent meets the six axioms of extents as follows.

1. The extent begins and ends with rounds since $((ab)^3ac)^3 = w^3 = e$.

2. No other row is repeated. (This is guaranteed by the coset decomposition of S_4.)

3. From one row to the next, no bell moves more than one position. (This is because $a = (12)(34)$, $b = (23)$, and $c = (34)$.)

4. No bell rests in the same place more than two successive rows. (The alternation of $a = (12)(34)$ meets this condition.)

5. The working bells (here all but the treble) do all the same work. (Since $w = (234)$, what bell 2 does in the first lead, bell 3 does in the second lead, and bell 4 does in the last lead.)

6. Each lead should be palindromic in its changes. $((ab)^3a = abababa.)$

It should be noted that White shows how the leads contain other cosets and the alternating group A_4. The reader should see [13, 10, 11].

2.5   Connections to a Cayley Graph

White also has proven a beautiful theorem [10] that relates the existence of an extent to the existence of a Hamiltonian cycle in a Cayley graph. Since most of my students have taken Discrete Algebraic Structures and are aware of the rudiments of graph theory, I usually include this in my course. A *Hamiltonian cycle* in a graph is a path that goes through every vertex exactly once, except for ending where it started. Let T be the set of all transitions that are employed for a given extent. T will then generate S_n. The *Cayley graph* $G_T(S_n)$ has as its vertex set S_n and as its edge set $E = \{\{g, tg\} \mid g \in S_n, t \in T\}$. Then the following theorem holds [10, 11, 12]:

Theorem 3 *An extent on n bells, using transition rules from T, can be composed if and only if $G_T(S_n)$ is Hamiltonian.*

For example, for Plain Bob Minimus, $T = \{(12)(34), (23), (34)\}$, which are our transition rules. The vertices of our Cayley graph are all the elements of S_4. An edge is drawn between two vertices if and only if one of the transition rules applied to one of the vertices give the second vertex. The Cayley graph for Plain Bob Minimus is given in Figure 1. A Hamiltonian circuit consists of the vertices listed in the same order as our extent. That is, 1234—2143—2413—4231–...–1342—3124–...–2134—1243—1234.

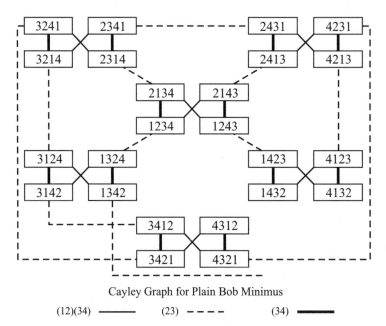

Cayley Graph for Plain Bob Minimus

(12)(34) ———   (23) – – – –   (34) ▬▬▬

Figure 1: Hamiltonian circuit for Plain Bob Minimus.

3 Putting Change Ringing into the Classroom

What we have seen thus far is only a small amount of the mathematics involved in change ringing. Since I teach a one-semester class in abstract algebra, I can only bring into the classroom the limited amount shown above. I do like to have my students experience change ringing for themselves. We happen to be located an hour from Old North Church in Boston where change ringing occurs every Saturday, so sometimes I am able to arrange a field trip. At a minimum, I like to play a tape of change ringing in the classroom, such as [7]. It is also fun to try to simulate change ringing in the classroom. Some people might be able to borrow hand bells from a church. However, bear in mind that hand bells are rather delicate, as well as being expensive. If one does not know how to train in the proper ringing technique, borrowing hand bells might not be a good idea. As a substitute, one can bring in an assortment of hand instruments or even homemade noisemakers. It would be more pleasing if the "instruments" at least somewhat mesh musically with each other. My classes always enjoy stumbling through some extents such as the Plain Bob Minimus. If all else fails, one can fall back on the method of training change ringers suggested in [14]: Find some stairs and have students practice changing places on the stairs following the appropriate extent.

I also suggest trying to get students to compose some extents on a reasonable number of bells. If one has access to a music sequencer program, one can have the students' efforts played back for them to hear. I happen to be a carillonneur, so I have been able to tape select performance of their work on a carillon. There is also a nice web site by Kees van den Doel at [9] that has a free, downloadable applet that plays change ringing compositions. My students enjoy hearing their compositions using that program.

The following are examples of questions I use as homework assignments.

1. Suppose you have four bells. Assume that $a = (12)(34)$, $b = (23)$, $c = (34)$, and $d = (12)$. Create the extent for a change ringing method using the multipliers w below. Write the extent both in permutation notation and with the bell numbers in permuted order. For example, if $w = (ab)^3ac$, you start your method with the identity permutation (1) or 1234. The next entry in your first lead is a, that is, $(12)(34)$ or 2143. The third entry of the lead is $(12)(34)$ multiplied *on the left* by b. So $(23)(12)(34) = (1342)$ or 2413. Since w has $(ab)^3$ as the beginning, our next entry is the old one multiplied *on the left* by a. So $(12)(34)(1342) = (14)$ or 4231. The next entry is the former entry multiplied on the left by b. Then you multiply that result on the left by a. Then by b, then by a, and finally by c. You have started your second lead at this point (with w). You now want to multiply each entry of the first lead (except the identity) *on the right* by all of w. Your third lead will be created by multiplying each entry of the second lead *on the right* by w. You should have 25 rows in your extent and the last one will also be (1) or 1234.

   (a) Double Bob: $w = (abadabac)$
   (b) Reverse Bob: $w = (abad(ab)^2)$
   (c) Reverse Canterbury: $w = (db(ab)^2dc)$

2. Go to the web site www.cs.ubc.ca/spider/kvdoel/bells/bells.html and input your leads for each method in question #1 following the directions at the site or on my handout. Don't forget to add rounds at the beginning. Listen to each of the change ringing methods you've written out. Which of the methods meet all of the six requirements of a modern extent? Why?

3. For each w in question #1, compute the cosets Hw, Hw^2, and Hw^3. Do the leads for any of the methods exactly fit the cosets, or are the cosets scrambled up among the leads? What do you think must be necessary for the leads to exactly fit the cosets as is the case for Plain Bob Minimus?

4. A palindrome is a word that is symmetric about the middle, such as *aba* or *abba* or *abacaba*. Notice that each word w in question #1 is a palindrome if the last letter of the word is omitted. Create your own word w with these properties that has a palindromic part of seven letters and an eighth letter based on the letters a, b, c. Make sure your word differs from the w's in question #1. Using the permutation values for a, b, c given in question #1, write out the three leads for the change ringing method you have just created. Be sure to list both the permutation notation and the bell notation in your leads. Which of the six rules for a modern extent does your method fulfill? Explain your observations. Go to the web site in question #2 and listen to your creation.

5. Draw a Cayley graph for your composition in question #4. Do you have a Hamiltonian cycle?

My students usually moan and groan when confronted with the exercises above. However, with a little help in getting started, they soon get fascinated by the problems. It is not unusual for my students to come to the next class after doing the homework and say, "Wow! That was *fun.*" Students report that playing back their compositions using the Java applet was the most enjoyable part. Some of the high school teachers decided to use the applet to play change ringing examples in their classes when introducing the concept of permutations.

4   A Final Note—Change Ringing and the American Revolution

My students are always fascinated to hear about the "backdoor" role change ringing played for Paul Revere's ride the evening of April 18, 1775. The English have always exported change ringing to their colonies, albeit not too successfully. The first ring of change ringing bells in the U.S. arrived at Boston in 1745 [2]. The eight bells were placed in Old North Church. Paul Revere was a parishioner of the church all of his life. So, shortly after the bells were hung, at age 10 young Paul Revere trained to be a ringer. The account books of the church showed that he was paid for ringing regularly for quite a few years. Thus, it is not surprising that when a place for signaling was needed that could be seen from across the Charles River, he and other Sons of Liberty had no trouble thinking of (and gaining access to) the tall tower of Old North Church.

After the American Revolution, Paul Revere became one of the first American bellfounders. In 1792 he started a bellfoundry with his son Joseph Warren. Together they cast 398 bells [2]. Many of these are still in existence up and down the East Coast.

References

[1] J. Camp, *Bell Ringing*, Latimer Trend & Co. Ltd., Gt. Britain, 1974.

[2] W. De Turk, *Meneely Bells: An American Heritage*, Bulletin of the Guild of Carillonneurs in North America **27** (1978), 30–61.

[3] R. Duckworth, *Tintinnalogia, or the art of ringing*, Kingsmead Reprints, Bath, 1970.

[4] R. Johnston, *Bell-ringing*, Viking, New York, 1986.

[5] E. Morris, *The History and Art of Change Ringing*, EP Publishing, Wakefield, 1974.

[6] D. Nowosielski, *Change Ringing: A Connection Between Mathematics and Music*, Pi Mu Epsilon **10** (1997) Fall, 532–539.

[7] Saydisc Records, *Church Bells of England*, Saydisc Records, Chipping Manor, The Chipping, Wotton-Under-Edge, Glos. GL12-7AD, England.

[8] F. Stedman, *Campanalogia: Or Art of Ringing Improved*, W. Godbid, London, 1677.

[9] K. van del Doel, web site, `http://www.cs.ubc.ca/spider/kvdoel/bells/bells.html`.

[10] A. White, *Ringing the Cosets*, American Math Monthly **94** (1987), 721–746.

[11] A. White, *Treble dodging minor methods: ringing the cosets, on six bells*, Discrete Mathematics **122** (1993), 307–323.

[12] A. White and R. Wilson, *The Hunting Group*, The Mathematical Gazette **79** (1995), 5–16.

[13] A. White, *Fabian Stedman: The First Group Theorist?*, American Math Monthly **103** (1996), 771–778.

[14] W. Wigram, *Change-Ringing disentangled: with hints on the direction of belfries, on the management of bells*, 2nd ed., London, England, 1880.

[15] W. Wilson, *Change Ringing: The Art and Science of Change Ringing on Church and Handbells*, October House, Inc., NY, 1965.

Lucy I. Dechéne, Department of Mathematics, Fitchburg State College, Fitchburg, MA 01420; `ldechene@fsc.edu`.

Part IV

Appendix

Internet Resources for this Volume

The reader may note that several of the articles in this volume refer to software, web sites, colored versions of diagrams, or other resources. Furthermore, several of the articles deal with on-going projects for which updates of materials may be available. To accommodate these various needs, we have created a web site dedicated to supplement this volume. It can be found using the following URL.

`http://www.central.edu/MAANotes/`

While the exact content and presentation of this site will change as the need arises, its basic structure will be maintained. After viewing an introduction and overview of the site, one is given a choice of viewing the preface, the list of abstracts, information related to the articles (organized by author, title, or section), short biographies of the authors, the list of software used, and a list of general abstract algebra links. Each article of this volume has its own page that includes the minimal information (title, author(s), institution(s), abstract, and a short biography) and, when appropriate, may also contains links to more information about the software used or other additional information that the author wishes to provide. The page listing the software used in this volume provides links to the publishers or creators of the software, information about the platforms under which the software runs, a link to a source for downloading (where appropriate), and when possible, links for further information about the software (tutorials, for example).

Some software information is provided in Table 1. For the most complete and accurate information, however, the reader is encouraged to use the web site established for this volume.

software	source
AbstractAlgebra	`http://www.central.edu/eaam`
Exploring Small Groups	accompanies *Laboratory Experiences in Group Theory*, from the MAA
Finite Group Behavior	`http://unr.edu/homepage/keppelma/fgb.html`
GAP	`http://www-gap.dcs.st-and.ac.uk/gap`
ISETL	
DOS	`http://archives.math.utk.edu/software/msdos/ discrete.math/isetl/isetl.zip`
Macintosh	`http://helios.tns.utk.edu/math_archives/ software/mac/progLanguages/ISETL/ISETL.sea.hqx`
Windows	`http://csis03.muc.edu/isetlw/isetlw.htm`
Maple	`http://www.maplesoft.com`
Mathematica	`http://www.wolfram.com`
MATLAB	`http://www.mathworks.com/`
PascalGT	`http://faculty.salisbury.edu/~kmshannon/pascal/pgtdown/welcome.htm`

Table 1: Sources for software.

About the Authors

Michael Bardzell is an Assistant Professor of Mathematics in the Department of Mathematics and Computer Science at Salisbury University. He received a B.S. in Physics from Mary Washington College in Fredericksburg, Virginia and his M.S. and Ph.D. in Mathematics from Virginia Polytechnic Institute and State University. His research interests are ring theory, symbolic computation, and cellular automata. He regularly attends the National Conference on Undergraduate Research. Several of his students have already presented work related to the PascGalois project.

sarah-marie belcastro is an Assistant Professor of Mathematics at the University of Northern Iowa. She specializes in algebraic geometry (surfaces that are hypersurfaces in toric varieties) and the mathematics of paper folding. After attending Haverford College as an undergraduate, she studied at the University of Michigan for her Ph.D., which she received in 1997. In addition to being a mathematician, sarah-marie is also a member of a dance company and thinks about feminist philosophy of science.

Steve Benson received his Ph.D. in 1988 from the University of Illinois, working under the direction of Leon McCulloh in algebraic number theory. Having held faculty positions at St. Olaf College, Santa Clara University, University of New Hampshire, and University of Wisconsin – Oshkosh, he is currently a Senior Research Associate at Education Development Center, in Newton, Massachusetts and a Co-Director of the Master of Science for Teachers program at the University of New Hampshire.

Ruth Berger is an Associate Professor of Mathematics at Luther College in Decorah, Iowa. She earned her Ph.D. in algebraic number theory from Louisiana State University in 1988 and then taught at Memphis State University for five years. In 1993 she came to Luther College where she enjoys the emphasis on good teaching and the student-oriented atmosphere.

Laurie Burton is an Assistant Professor of Mathematics and Mathematics Education at Western Oregon University where she transferred after teaching for four years at Central Washington University. She is trained as a commutative ring theorist and is currently involved in the mathematics education of elementary and middle school teachers. She enjoys designing classes that utilize different educational approaches to enhance active and visual student learning. After attending the California State University at Chico as an undergraduate, she studied at the University of Oregon for her Ph.D., which she received in 1995.

Kevin Charlwood is an Assistant Professor in the Department of Mathematics and Statistics at Washburn University in Topeka, Kansas. He earned his Ph.D. from the University of Wisconsin - Milwaukee in quantum groups in 1994. His current research interests lie in using technology to enhance student learning in calculus and modern algebra.

Lucy Dechéne is a Professor of Mathematics at Fitchburg State College in Fitchburg, Massachusetts. She received her B.S. in mathematics (with a second major in organ performance) from the University of San Francisco and her M.S. and Ph.D. degrees from the University of California, Riverside. Her activities also include being a professional organist, carillonneur, and composer. When teaching, she enjoys combining her love of music with her love of mathematics.

Suzanne Dorée is an Associate Professor and Chair of the Mathematics Department at Augsburg College in Minneapolis, Minnesota, where she has taught since 1989. She earned her Ph.D. in character theory from the University of Wisconsin - Madison. Her research interests include curriculum and materials development and directing undergraduate student research. She enjoys teaching mathematics at all levels, using pedagogies that support active learning, and teaching mathematical thinking, writing, and speaking skills.

Brad Findell earned an M.A. in mathematics from Boston University in 1990 and recently completed his Ph.D. in mathematics education at the University of New Hampshire under the direction of Joan Ferrini-Mundy and Karen

Graham. He is now in the Mathematics Education Department at the University of Georgia. From 1997 to 2001, he was at the National Research Council in Washington, DC, working for the Mathematical Sciences Education Board (MSEB) and on various projects in mathematics education. He served as editor of *High School Mathematics at Work* (MSEB, 1998), and is coeditor, with Jeremy Kilpatrick and Jane Swafford, of *Adding It Up: Helping Children Learn Mathematics* (2001), both published by the National Academy Press.

Paul Fjelstad helped design St. Olaf's Paracollege, which started in 1969 with the goal of encouraging students to take the initiative in designing their education and to experiment with learning styles in the process. His graduate work was in physics (Harvard Ph.D., 1962), where the concrete and abstract work on each other intermittently. In interacting with students, he prefers asking questions to giving answers.

Gary Gordon, a native Floridian, received his B.A. from the University of Florida in 1977 and his Ph.D. from the University of North Carolina in 1983. He has taught mathematics at Lafayette College since 1986, and before that, at Williams College. He has also worked in industry. His mathematical interests include combinatorics, geometry, and algebra. Most of his research has involved analyzing polynomial and other invariants for graphs, posets, and other combinatorial objects. He loves watching baseball and playing softball, tennis, golf, and other sports where people swing clubs. He also enjoys all sorts of games, but usually loses to his wife and frequent mathematical collaborator, Liz McMahon, and to his two daughters, Rebecca and Hannah.

Al Hibbard is Professor of Mathematics at Central College in Pella, Iowa. He did his undergraduate work at St. John's University (MN) and obtained his Ph.D. from the University of Notre Dame. Since 1990 he has been investigating various ways of incorporating technology and other innovations in the teaching of abstract algebra (and other courses). One outgrowth of these pursuits is that he and Ken Levasseur have created a suite of *Mathematica* packages (and lab notebooks based on them) that enable one to work with groups, rings, and morphisms. These *AbstractAlgebra* packages are the foundation of their book *Exploring Abstract Algebra with Mathematica*. When not teaching, programming, or writing, he can be found in a "committee meeting" in a racquetball court at the athletic facilities.

Ed Keppelmann is an Associate Professor of Mathematics at the University of Nevada in Reno. He is an active member of the MAA and does research in Nielsen fixed point and periodic point theory.

John O. Kiltinen, Professor of Mathematics at Northern Michigan University, earned his Ph.D. in topological algebra at Duke University in 1967 and then taught four years at the University of Minnesota. He was a Fulbright lecturer at the University of Joensuu and a visiting professor at the Tampere University of Technology in 1978–79. As the first acting director of NMU's Glenn T. Seaborg Center for Teaching and Learning Science and Mathematics, he directed the Michigan Mathematics Early Placement Test and several other grant-funded projects. His research has been in the areas of topological and pure algebra.

George Mackiw is Professor of Mathematical Sciences at Loyola College in Baltimore, Maryland. He did his undergraduate work at Georgetown University, Washington, D.C. and received his Ph.D. from the University of Virginia. In 1996 he was presented the Award for Distinguished College or University Teaching by the Maryland/D.C./Virginia Section of the Mathematical Association of America. His interests include abstract algebra, coding theory, linear algebra, and working with students.

Ellen Maycock (formerly Ellen Maycock Parker) studied group theory as an undergraduate at Wellesley College and returned to this early interest when she wrote *Laboratory Experiences in Group Theory*, published by the Mathematical Association of America in 1996. She studied algebraic topology as a graduate student and completed an M.S. and Ph.D. at Purdue University, writing a dissertation in the area of operator algebras. She taught at Purdue, Indiana University-Purdue University at Indianapolis, and Wellesley before joining the faculty of DePauw University, where she is now Professor of Mathematics. She is energized by teaching interdisciplinary courses and is currently working on a non-mathematical book entitled *In Search of Her Mother's Paris*. She enjoys being the mother of two children, and in-line skates for fun and exercise.

Moira McDermott is an Assistant Professor of Mathematics at Gustavus Adolphus College. She specializes in commutative algebra and is particularly interested in characteristic p methods and computational algebra. After attending Bryn Mawr College as an undergraduate, she studied at the University of Michigan for her Ph.D., which she received in 1996. She was a Visiting Assistant Professor at Bowdoin College for two years before joining the faculty of Gustavus.

Karin M. Pringle is an Associate Professor in the Department of Mathematics and Statistics at the University of North Carolina at Wilmington. She received her doctorate of philosophy in mathematics at the University of Oregon in 1990.

Julianne Rainbolt is an Assistant Professor in the Department of Mathematics and Mathematical Computer Science at Saint Louis University in Saint Louis, Missouri. She received a B.A. in mathematics and philosophy from Ohio Wesleyan University. She earned her Ph.D. in group representation theory from the University of Illinois at Chicago in 1996. From 1996-1998 Dr. Rainbolt held a postdoctoral position at Michigan State University before her current position at Saint Louis University.

Kathleen Shannon is a Professor of Mathematics and Chair of the Department of Mathematics and Computer Science at Salisbury University. She received her B.A. in Mathematics and Physics from the College of the Holy Cross and her M.Sc. and Ph.D. in Applied Mathematics from Brown University. She is a student chapter coordinator for the Mathematical Association of America's Maryland/D.C./Virginia Section. Her interests include humanistic mathematics, scientific computing, approximation theory, general education, and mathematics education.

Robert Smith is a Professor of Mathematics and Statistics at Miami University and the Director of the Junior Scholars Program, a pre-college program for high school students. He received a B.S. from Morgan State College in 1963 and graduate degrees from Pennsylvania State University in 1967 and 1969. While he has been at Miami since 1969, his career has been punctuated with visiting positions in the U.S. and Australia. Since 1987, he has been involved in numerous innovative computer-based teaching projects. Perhaps the most fascinating of these is teaching abstract algebra using ISETL (a mathematical programming language) and cooperative learning.

Bayard Webb is the creator of *Finite Group Behavior*. He has a B.S. and M.S. in mathematics from the University of Nevada Reno. He currently works for International Game Technology where he designs slot machines and other games for casinos around the world.

John Wilson thoroughly enjoys explaining mathematics to people of any age. He has tutored or taught mathematics every year since he was in the seventh grade. After receiving his Bachelor of Science degree at the University of the South he earned his M.S. and Ph.D. at the University of North Carolina. He has taught at Centre College in Danville, Kentucky since 1985. He married his high school sweetheart and they have two children. He enjoys playing games and solving puzzles of all types but receives a special pleasure when they lead him to an interesting mathematical question.

Index

AbstractAlgebra, 97, 145
active learning, 4
addition (mod 2), 19
algebra, 20
assessment, 8
authors, information about, 147
automorphisms, 68

Cayley graph, 30, 140
change, 137
change ringing, 137
classroom atmosphere, 13, 74
collaboration, 5
commutators, 135
complex numbers, 36, 95
conjectures, 78
conjugation, 28, 135
constructivism, 11, 73
cooperative learning, 72
cosets, 30, 59, 66, 102, 139
course design, 12
crystallographic groups, 29
cycles, 92, 108, 132

dihedral group, 87, 119
Dirac notation, 24
direct isometry, 27
direct product, 81, 107

equation, solution of, 36
Euclidean group, 25
exams, take-home, 8
exploratory exercises, 6, 19, 77, 89, 106
Exploring Small Groups, 12, 42, 45, 56, 63, 145

factor groups, *see* quotient groups
factorials, application of, 138
finite field, 85, 126, 127
Finite Group Behavior, 45, 145
first-day activity, 5, 12
fractals, 116
frieze groups, 30

GAP, *see* Groups, Algorithms and Programming
generators and relations, 29
Geometer's Sketchpad, 26

$GL(n, p)$, 85
glide reflection, 27
group properties, 58, 99
group work, 5, 72
`Groupoid`, 99
groups
    crystallographic, 29
    dihedral, 87, 119
    Euclidean, 25
    frieze, 30
    matrix, 85, 93
    modular, 65
    permutation, 65, 131, 137
    symmetry, 92
Groups, Algorithms and Programming, 77, 145

Hamiltonian cycle, 140
homework
    collaborative, 6
    feedback, 8
    keep in mind when writing, 4
    rewriting, 8
homomorphisms, 50, 59, 61, 104, 108
Hungarian Rings puzzle, 134
hunting group, 139
Hypercard, 131

idempotent, 80
indirect isometry, 27
ISETL, 55, 63, 71, 74, 145
isometry, 25
    direct, 27
    indirect, 27
    of the plane, 27
isomorphisms, 68, 108

KaleidoTile, 32
Kali, 32

labs, using, 42, 55, 98
Lagrange's Theorem, 59
lecturing, avoiding, 5
Lights Out, 125
linear algebra, 126

MAGMA, 46

Maple, 91, 145
Mathematica, 97, 145
MATLAB, 85, 145
matrix groups, 85, 93
MetaCard, 131
modular arithmetic, 12, 36, 65, 74, 80, 115
modular groups, 65
Morphoid, 99

nilpotent, 80
normal subgroup, 30, 102

order
    of a group, 107
    of an element, 82, 86, 106
    of matrix groups, 86
Oval Track puzzle, 132

p-group, 119
paper, algebraic application of, 20
papers, student, 7
Parity Theorem, 135
participation, 5, 6
Pascal's triangle, 115
PascalGT, 116, 145
PascGalois Triangle, 116
penny, application of, 19
permutation groups, 65, 131, 137
permutations, 92, 132, 139
Plain Bob Minimus, 138
pod, an algebraic system, 20
Project NExT, 3
projects
    group, 125
    planning, 125
    research, 35, 42
    term, 7
proofs
    analysis, 6
    writing, 4
puzzles, 125, 131

quaternion group, 88
quotient group, 30
quotient groups, 15, 50, 67, 102, 118

reflection, 27
research questions, 36
rewriting homework, 8
Ringoid, 99
rings, 103
role of technology, 43
roots of unity, 36, 95
rotation, 27

self-similarity, 119
semiring, 23
$SL(n,p)$, 86
Slide puzzle, 133
software sources, 145
solutions to an equation, 36
student papers, 7
subgroups, 49, 107
syllabus, 4
Sylow theorems, 51
symmetry groups, 92

term projects, 7
TesselMania, 32
text, use of a, 13
Top Spin, 131
tracks, an algebraic system, 21
translation, 27
Transpose puzzle, 132
transpositions, 132, 138

Unique Factorization Domain, 110

vector space, 24
visualization, 42, 48, 99, 101

web site for this volume, 145
writing
    homework, 4
    pointers, 4
    proofs, 4
    reports, 37, 41